Topics in Recreational Mathematics 3/2016

Editor-in-chief

Charles Ashbacher
5530 Kacena Ave
Marion, IA 52302 USA

cashbacher@yahoo.com

Artwork

Caytie Ribble

Problems

Lamarr Widmer

Contributor

Rachel Pollari

ISBN-13: 978-1537333212

Copyright 2016

CONTENTS

Note From the Editor — 5
Charles Ashbacher

Mathematical Cartoons — 6
Caytie Ribble

Nonamorphic Numbers Revisited — 8
Charles Ashbacher

Octamorphic Numbers Revisited — 10
Charles Ashbacher

Pentamorphic Numbers Revisited — 13
Charles Ashbacher

Mathematical Kabobs — 15
Charles W. Trigg and Charles Ashbacher

African Horizons: A Math Enrichment Experience in Kenya — 17
Lamarr Widmer

Elementary Problems — 20

The Life-Time of the World — 22
A.A.K. Majumdar

WordPlay — 26
Rachel Pollari and Charles Ashbacher

Mathematicians in the "School of Athens" — 27
Charles Ashbacher

Solutions to Mathematical Kabobs — 29

More Irreverence? — 30
Charles Ashbacher

Palindromic Numbers and Iterations of the Pseudo-Smarandache Function — 32
Charles Ashbacher

Divisibility and Periodicity Patterns and Palatable Number Tricks in the Jacobsthal Sequence — 35
Jay Schiffman

Abstracts of the papers in "Journal of Recreational Mathematics" Volume 1, Number 1, 1968 — 48
Charles Ashbacher

Smarandache Bisymmetric Geometric Determinant Sequence — 50
A.A.K. Majumdar

Solutions to Elementary Problems — 60

The NFL Draft, 2002-2014: Winners, Losers, and a New Draft Trade Value Chart — 61
Paul M. Sommers

Wordplay Sayings Translated — 80

Solution to Mathematicians in the "School of Athens" — 81

Book Reviews — 82
edited by Charles Ashbacher

Solutions Column, Journal of Recreational Mathematics Problems 2880 – 2889 From 38(2) — 87
edited by Lamarr Widmer

Proposers And Solvers List For Problems And Conjectures Journal of Recreational Mathematics 38(2) — 98

Problems and Conjectures — 99
edited by Lamarr Widmer

Solutions To Problems From Topics in Recreational Mathematics 3/2015 — 101
edited by Lamarr Widmer

In the Loop—the Mini and Max of Paths — 108
by Kate Jones

Neutrosphic Sets and Systems 115

**Books in Recreational Mathematics by
Charles Ashbacher and Associates** 117

Note From the Editor

Hello and welcome to the third **Topics in Recreational Mathematics** book of 2016 and the eighth in the series. Our goal in producing this series is to keep the flame of recreational mathematics going, it has so much to offer the world in terms of quality entertainment and mathematical education.

Towards that end, I remain active in creating additional works derived from the 38 volumes of **Journal of Recreational Mathematics**. A book containing all the table of contents pages from all 38 volumes is either already out or will be shortly.

Work also is moving forward on creating a collection of abstracts to all the papers that were published in **JRM**. It is my goal to keep the memory of **JRM** alive as so much quality and entertaining mathematics appeared within the pages. One of the entries in this collection is the abstracts to the papers that appeared in **JRM** volume 1, number 1. Since these papers appeared nearly fifty years ago and none had abstracts when published, all of them were written by me.

Many years ago, Joe Madachy, the longtime editor of **JRM** told me that by far, Charles W. Trigg was the most prolific contributor to **JRM**, a quick scan of the first 25 volumes will quickly convince you of that. Therefore, I thought it fitting that some of his work be revisited and expanded. Some short items in this collection do just that.

There is also another section devoted to wordplay authored by Rachel Pollari. I find such material to be entertaining and relaxing, in a word, recreational.

Lamarr Widmer, the problem editor, spent the 2009-2010 academic year at Daystar University in Nairobi, Kenya. In a short item, he describes how he used a recreational mathematics club to improve the quality of the interaction he had with the students.

As always, I welcome feedback of all kinds as well as contributions. Of special interest are papers that expand on material that appeared in **JRM** as well as undergraduate research projects. Number nine in the **TRM** series is already in production.

Charles Ashbacher

cashbacher@yahoo.com

Mathematical Cartoons

Caytie Ribble

THE PROPER WRAP TO A MATH PRESENT

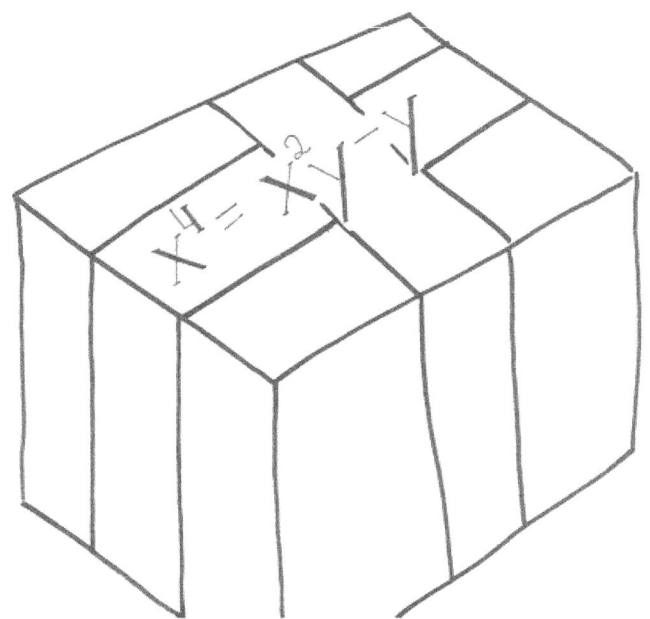

Common error made by topologists

Nonamorphic Numbers Revisited

Charles Ashbacher
cashbacher@yahoo.com

Abstract

In volume 20, number 2 of **Journal of Recreational Mathematics**, Charles W. Trigg defined a **nonamorphic number** to be a nonagonal number $N(n) = n(7n - 5) / 2$ that terminates in n. For example, $N(25) = 2125$ and $N(625) = 1365625$. The purpose of this paper is to report the results of a greater search for nonamorphic numbers.

In volume 20, number 2 of **Journal of Recreational Mathematics**, Charles W. Trigg started with the nonagonal numbers defined by the formula $N(n) = n(7n - 5) / 2$ and defined the **nonamorphic** numbers to be nonagonal numbers that terminate with the index of the nonagonal number.

Trigg identified five base-ten nonamorphic numbers less than 10^4 and they are: $N(1) = 1$, $N(5) = 75$, $N(13) = 1113$, $N(625) = 1365625$ and $N(9376) = 307659376$.

I recently encountered this paper when I was reading through old issues of **JRM** and wondered if any additional work had been done in this area. Naturally, I did a Google search but there was no mention of the nonamorphic numbers as defined by Trigg. Therefore, using the BigInteger class in the progamming language Java, I conducted a more extensive search for additional nonamorphic numbers in base ten.

The search was extended up to 10^9 and five additional nonamorphic numbers were found. They appear in table 1.

Table 1

n	N(n)
90625	28744890625
890625	2776242890625
7109376	176901277109376
12890625	581588712890625
212890625	158628463212890625

The most obvious point of interest is the values of the last three and four digits, which are repeats of the digits of the larger two of the nonamorphic numbers found by Trigg. This leads to three additional unresolved questions.

1. Are there additional nonamorphic numbers?

2. If there are more nonamorphic numbers, do they have the same trailing digits as those in table 1 or are they different?

3. After the first two, there is only one nonamorphic number for each number of digits in n. Does this pattern continue?

Reference

1. C. W. Trigg, *Nonamorphic Numbers*, **Journal of Recreational Mathematics**, 20(2). pp. 97-98, 1988.

Octamorphic Numbers Revisited

Charles Ashbacher

cashbacher@yahoo.com

Abstract

In volume 19, number 2 of **Journal of Recreational Mathematics**, Charles W. Trigg defined an **octamorphic number** to be an octagonal number $E(n) = n(3n - 2)$ that terminates in n. For example, $E(25) = 1825$ and $E(625) = 1170625$. The purpose of this paper is to report the results of a greater search for octamorphic numbers.

In volume 19, number 2 of **Journal of Recreational Mathematics**, Charles W. Trigg started with the octagonal numbers defined by the formula $E(n) = n(3n – 2)$ and defined the **octamorphic** numbers to be octagonal numbers that terminate with the index of the octagonal number.

Trigg identified eight base-ten octamorphic numbers less than 10^5 and they are: $E(1) = 1$, $E(5) = 65$, $E(6) = 96$, $E(25) = 1825$, $E(76) = 17176$, $E(376) = 423376$, and $E(9376) = 263709376$.

I recently encountered this paper when I was reading through old issues of **JRM** and wondered if any additional work had been done in this area. Naturally, I did a Google search but there was no mention of the octamorphic numbers as defined by Trigg. Therefore, using the BigInteger class in the progamming language Java, I conducted a more extensive search for additional octamorphic numbers in base ten.

The search was extended up to 10^9 and additional octamorphic numbers were found. They appear in table 1.

Table 1

n	E(n)
90625	24638490625
109376	35889109376
890625	2379636890625
2890625	25067132890625
7109376	151629667109376
12890625	498504612890625
87109376	22764129987109376
787109376	1858623507787109376

The most obvious point of interest is the values of the last three and four digits, which are repeats of the digits of the larger two of the octamorphic numbers found by Trigg. This naturally leads to two additional unresolved questions.

1. Are there additional octamorphic numbers?

2. If there are more octamorphic numbers, do they have the same trailing digits as those in table 1 or are they different?

The trailing digits of the values of n of the known octamorphic numbers indicate the potential for an infinite family.

Reference

1. C. W. Trigg, *Octamorphic Numbers*, **Journal of Recreational Mathematics**, 19(2). pp. 116-118, 1987.

Pentamorphic Numbers Revisited

Charles Ashbacher
cashbacher@yahoo.com

Abstract

In volume 16, number 1 of **Journal of Recreational Mathematics**, Charles W. Trigg defined a **pentamorphic number** to be a pentagonal number $P(n) = n(3n - 1) / 2$ that terminates in n. For example, $P(25) = 925$ and $P(625) = 585625$. The purpose of this paper is to report the results of a greater search for pentamorphic numbers.

In volume 16, number 1 of **Journal of Recreational Mathematics**, Charles W. Trigg started with the pentagonal numbers defined by the formula $P(n) = n(3n - 1) / 2$ and defined the **pentamorphic** numbers to be pentagonal numbers that terminate with the index of the pentagonal number.

Trigg identified three non-trivial base-ten pentamorphic numbers less than and they are: $P(1) = 1$, $P(5) = 35$, $P(25) = 925$ and $P(625) = 585625$. Trigg specifically states that there are no other pentamorphic numbers in base 10.

I recently encountered this paper when I was reading through old issues of **JRM** and wondered if any additional work had been done in this area. Naturally, I did a Google search but there was no mention of the pentamorphic numbers as defined by Trigg. Therefore, using the BigInteger class in the progamming language Java, I conducted a more extensive search for additional pentamorphic numbers in base ten.

The search was extended up to 5×10^8 and additional pentamorphic numbers were found. They appear in table 1.

Table 1

n	P(n)
9376	131859376
90625	12319290625
890625	1189818890625
7109376	75814837109376
12890625	249252312890625
212890625	67983627212890625

The most obvious point of interest is the values of the last three and four digits, which are repeats of the digits of the largest of the pentamorphic numbers found by Trigg. This leads to two additional unresolved questions.

1. Are there additional pentamorphic numbers?

2. If there are more pentamorphic numbers, do they have the same trailing digits as those in table 1 or are they different?

The trailing digits of the values of n of the known pentamorphic numbers indicate the potential for an infinite family.

Reference

1. C. W. Trigg, *Pentamorphic Numbers*, **Journal of Recreational Mathematics**, 16(1). pp. 116-118, 1987.

Mathematical Kabobs

Charles W. Trigg

Charles Ashbacher

cashbacher@yahoo.com

In Volume 19, number 1 of **Journal of Recreational Mathematics**, Charles W. Trigg defined a problem that he called "An Arithmetic Kabob." That problem is reproduced here.

In each line, fill in the blanks with letters to form a term frequently encountered in arithmetic operations.

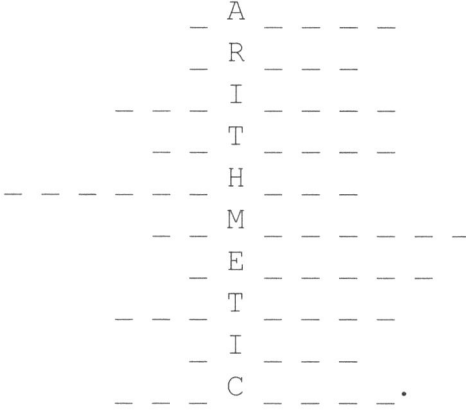

The solution appeared on a later page in that issue and is reproduced here.

```
      FACTOR
       PRIME
     DIVIDEND
      INTEGER
   SUBTRAHEND
     NUMERATOR
     DECIMAL
      QUOTIENT
        DIGIT
      FRACTION
```

Continuing this tradition and paying tribute to Trigg, three additional kabob problems have been created. The solutions to all three appear later in this issue.

Algebra Kabob

In each line, fill in the blanks with letters to form a term frequently encountered in algebra operations.

```
        _ _ _ _ _ A _ _ _ _
            _ _ _ _ L _ _ _
              _ _ G _ _ _ _ _
              _ _ E _ _ _ _ _
        _ _ _ _ _ B _ _ _ _ _
            _ _ _ _ R _ _
        _ _ _ _ _ A _ _ _ _
```

Geometry Kabob

In each line, fill in the blanks with letters to form a term frequently encountered in algebra operations.

```
        _ _ G _ _
          _ E _ _ _ _ _
          _ O _ _ _
        _ _ _ M _ _ _
          _ E _ _ _ _ _ _ _
            T _ _ _ _ _ _ _
        _ _ _ _ R _
            Y _ _ _ _ _
```

Numbers Kabob

In each line, fill in the blanks with letters to form a term that is a set of numbers.

```
        _ _ _ _ N _ _
          _ U _ _ _ _ _
        _ _ M _ _ _ _
            B _ _ _ _ _
          _ E _ _ _
        _ _ _ _ R _ _
        _ _ _ _ _ _ _ S
```

African Horizons: A Math Enrichment Experience in Kenya

Lamarr Widmer
Messiah College
widmer@messiah.edu

Abstract

The author's most recent teaching experience in Africa was one year of teaching mathematics at Daystar University. This paper describes challenges with the classroom culture at that institution and the author's attempt to motivate students through the use of mathematical activity of a more recreational nature.

I spent the 2009-2010 academic year at Daystar University in Nairobi, Kenya. I taught six different mathematics classes, including Calculus, Linear Algebra and Discrete Mathematics. I confronted a number of challenges related to my students' habits and expectations based on their previous experience. In the typical African classroom setting, the teacher's primary function is to deliver information, generally by lecturing. Assessment consists mostly of traditional exams which measure the student's level of retention of the information communicated in class. Problem-solving, critical thinking and research projects, while not unknown, are infrequently used.

The classroom culture does not encourage student participation. Students expect to spend their time dutifully recording the information provided by the lecturer. They are reticent to speak in class. I believe their reluctance to ask a question is often due to a desire to avoid betraying a lack of understanding. Teachers may communicate an impression that a rigid lesson plan would be disrupted by unexpected student questions or input. Cultural expectations of proper respect and deference engender a more distant student-teacher relationship than I expect in my American classroom. I announced office hours but had very few takers.

This situation was problematic as it runs counter to my desire for a more informal, collaborative classroom where I teach concepts and problem-solving skills as well as theorems, formulas and methods. I want to get to know my students and help them to experience learning as exploration. My classroom needs to be a place for two-way communication, with continual, voluntary student feedback which lets me discern the success or failure of our learning enterprise and adjust accordingly. This is an ideal requiring a level of comfort which takes time to establish between the student and an unfamiliar teacher.

In concert with my attempts to set a different tone within the classroom, I offered my students a one hour recreational mathematics club experience once per week. I emphasized that it was optional, not necessarily related to content or performance in any of my classes and that it would feature mathematics which I find particularly enjoyable. During these meetings I introduced various topics, including graph theory and number theory. I distributed copies of **Math Horizons**, a big hit with the students, and then discussed the content with them after they had a chance to read it. In one session, we used calculators to carry out iteration of the function $f(x) = \cos x$ and explore the idea of fixed points.

A highlight of the club was the day when we constructed Conway pencil models, a hands-on exercise in three-dimensional geometry which I learned in a workshop during an MAA section meeting. Each student left that session with the model he had constructed, a fine starter of conversations with fellow students. Near the end of the school year, I gave a talk about David Hilbert which included reference to the German political situation and his famous agenda for twentieth-century mathematical research. I also noted his influence as a teacher, using the

Mathematical Genealogy website to document his numerous "descendants." I was pleasantly surprised by my students' level of interest and appreciation for this historical background. So before the semester came to an end, I squeezed in a talk about the mathematics of ancient Egypt, a topic chosen partly for Egypt's proximity to Kenya.

The positive outcomes of this initiative were well worth the effort. A small group of African students, some destined to be teachers, are now aware of recreational mathematics and possess sample copies of **Math Horizons** and **Journal of Recreational Mathematics**, which I left with them. They have been introduced to the MAA and its services, including the **Horizons** website. I did get to know the students who attended my math club as they were more talkative in this setting. I am convinced that, for those who participated in the math club, this openness carried over to my classes. During the second semester I began to receive more students during my office hours, including a number of those from the math club.

This recreational math club activity contributed positively to my teaching mission at Daystar University. It proved to be a more conducive venue than the classroom for communicating my enthusiasm for mathematics. I believe that students saw me in a different light in this setting. They could appreciate my willingness to give my time to this activity which was voluntary on my part as well as theirs. **Math Horizons** and **Journal of Recreational Mathematics** were a crucial element in this success.

My teaching assignment at Daystar University was limited to only two semesters. This was sufficient time to understand and adapt to the classroom culture. The activity described above was my attempt to bring a different feel to the student-teacher relationship and a more engaging view of my discipline. I did enjoy a reasonable degree of autonomy in this situation, enough to carry out the next step in this experiment, if given another opportunity. This step would be to bring some of this same activity to the regular classroom where it would reach all students.

Elementary Problems

Charles Ashbacher

cashbacher@yahoo.com

This section contains a few problems that are elementary in nature. They are both old and new.

The following was contributed by Miklos N Szilagyi.

Solve the following puzzle, where each letter represents a unique digit.

```
              AEHB
        ┌─────────
     GH │  ABCDEF
           GH
           ──
           DED
           CJK
           ───
            FDE
            FBD
            ───
             BAF
             AEF
             ───
              FK
```

The following alphametic is based on episode 126 of "Star Trek: The Next Generation, Time's Arrow I." It opens with the head of DATA being found in a deep mine that has been closed for centuries. It is to be solved in base 10, the question mark is not part of the problem and of course DATA is the greatest.

```
    126
   DOES
   DATA
    DIE?
   TIME
   ─────
  LOOPS
```

The following alphametic is based on episode 127 of "Star Trek: The Next Generation, Time's Arrow II." It continues the story of DATA's head being found deep in an abandoned mine and how it is resolved with the help of one of the most famous American writers.

```
          1   2   7
    W   I   T   H
        T   H   E
    H   E   L   P
            O   F
    ─────────────────
T   W   A   I   N
```
Solve in base 12 and he gave the greatest HELP.

This puzzle appeared in the first issue of **Recreational Mathematics Magazine**.

How Spirited Are You?

Of three members of a club, two drink wine, two liquor and two beer. The one that does not drink beer does not touch liquor, and the one that does not drink liquor does not drink wine. Which of the three beverages do they drink, respectively?

The Life-Time of the World

A.A.K. Majumdar
APU, 1-1 Jumonjibaru, Beppu-shi 874-8577, Japan
majumdar@apu.ac.jp

Abstract

The legend is that, at the creation of the world, there was a Tower of Hanoi with three poles and 64 discs, in a temple. The priests there are in the process of transferring the tower from one pole to another, in minimum number of moves, where each move transfers one disc from one pole to another under the "divine rule" that no disc can ever be placed on top of a smaller one. As soon as the task of the priests would be completed, the world would come to an end. This paper examines the different cases when a single relaxation of the "divine rule" is allowed.

Keywords. The Tower of Hanoi, divine rule, the life-time of the world.

Introduction

In 1885, the famous French number theorist, Francois Edouard Anatole Lucas[1] (1842 – 1891), introduced a new mathematical puzzle in Europe, which became very popular during that time. The toy puzzle is as follows : Given are three pegs, S, P and D, and 8 discs of different sizes. Initially, the discs rest on the peg, S, in a *tower* in small-on-large ordering (with the largest disc at the bottom, the second largest above it, and so on, and the smallest disc at the top), as shown in figure 1. The problem is to transfer this tower from the peg S to the peg D, in original ordering, in minimum number of moves, where each move can transfer the topmost disc from one peg to another under the "divine rule" that no disc can ever be placed on top of a smaller one.

Figure 1

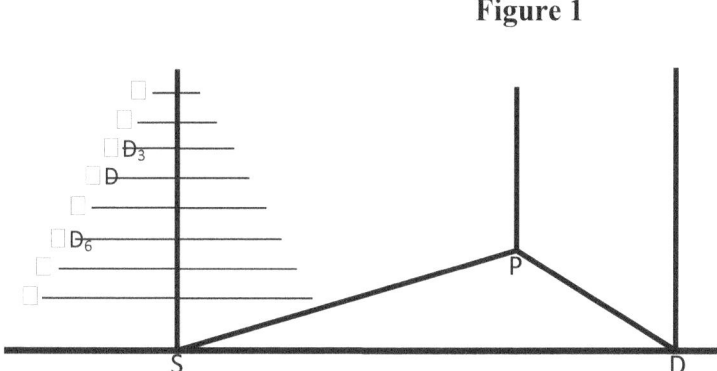

To market the puzzle, the following legend was attached to it (see, Ball[2] and Gardner[3]) :
"In the great temple of Benares beneath the dome, which marks the center of the world, rests a brass plate in which are fixed three diamond needles, each a cubit (18" – 22", or 45 – 56 cm) high and as thick as the body of a bee. On one of these needles, at the creation, God placed 64 discs of pure gold, in a tower, the largest disc resting on the brass plate, and the others getting smaller and smaller up to the top one. This is the *Tower of Brahma*. Day and night unceasingly, the priests transfer the discs from one diamond needle to another according to the fixed and immutable laws of Brahma, which require that the priest on duty must not move more than one disc at a time and that he must place this disc on a needle so that there is no smaller disc below it. When all the 64 discs shall have been transferred from the needle on which God placed them to one of the other needles, tower, temple and Brahmins alike will crumble into dust, and with a thunderclap the world would vanish".

In the more general setting, the problem, called the *Tower of Hanoi problem*, is as follows : Given are three pegs, S, P and D, and n ($\geq$ 1) discs of different radii, D_1, D_2, …, D_n, in increasing order. Initially, the discs rest on the *source peg*, S, in a *tower* in small-on-large ordering. This is illustrated in figure 2.

The objective is to transfer this tower from the peg, S, to the *destination peg*, D, in minimum number of moves (using the available pegs), where each move can transfer one (the topmost) disc from one peg to another, under the "divine" rule that no disc can ever be placed on a smaller one.

Figure 2

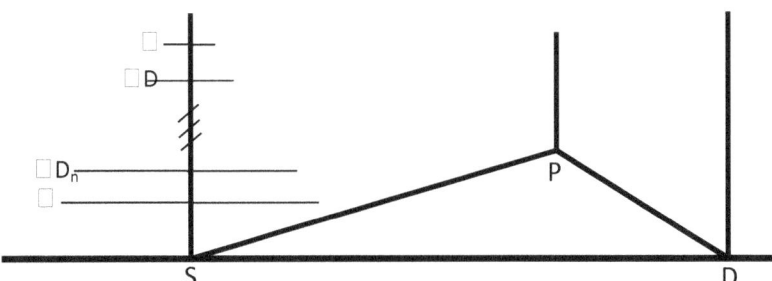

Let M(n) denote the minimum number of legal moves required to solve the above problem. Then, M(n) satisfies the following recurrence relation :

Theorem 1.1 : For all $n \geq 1$, $M(n) = 2M(n-1) + 1$; $M(0) = 0$.

The solution of the recurrence relation in Theorem 1.1 is given below.

Theorem 1.2 : $M(n) = 2^n - 1$, $n \geq 1$.

The Life-time of the World

Using Theorem 1.2, one can estimate the age of the world if one knows the present configuration of the tower in the temple. But how can we know it!

However, we can estimate the life-time of the world as follows : Since

$$M(64) = 2^{64} - 1 = 18,446,744,073,709,551,615,$$

assuming that each move takes 1 second, the priests of the Temple would require more than 5.84×10^{11} years to complete their assigned task.

But what happens if God, annoyed with the present situation of mankind, in a Sermon, allows the priests one relaxation of the "divine rule", so that, during the process of transferring the 64 discs from the source peg to the destination peg, for only one time the priests can break the "divine rule" by putting a larger disc on a smaller one.

In the more general setting, the modified problem is as follows : Given the three pegs, S, P and D, and n (≥ 1) discs of different sizes, resting on the source peg S in a tower, the problem is to shift the tower from the peg S to the peg D, in minimum number of moves, where each move can transfer only the topmost disc from one peg to another such that no disc is ever placed on top of a smaller one, except once. Let S(n) denote the minimum number of moves required to solve the modified problem. Then, we have the following result, due to Chen, Tian and Wang [4], giving an expression for S(n).

Theorem 2.1 : For any $n \geq 1$,
$$S(n) = \begin{cases} 2n - 1, & \text{if } 1 \leq n \leq 3 \\ 2^{n-2} + 5, & \text{if } n \geq 4 \end{cases}$$

For the problem with n (≥ 4) discs and one relaxation of the "divine rule", the optimal policy is as follows : First, transfer the tower of the topmost n – 3 discs D_1, D_2, …, D_{n-3} (from the peg S) to the peg P (in $2^{n-3} - 1$ moves), next shift the disc D_{n-2} to the peg D, then move the disc D_{n-1} (from the peg S) to the peg P on top of the tower of n – 3 discs (violating the "divine rule"), followed by the transfer of the disc D_{n-2} (from the peg D) to the peg P. After transferring the largest disc to the peg D, first shift the disc D_{n-2} (from P) to S, then move the disc D_{n-1} (from P) to D, next transfer the disc D_{n-2} (from S) to D and finally, move the tower of n – 3 discs (from P) to D to complete the tower on the peg D. Thus, the total number of moves involved is

$$2[(2^{n-3} - 1) + 1 + 1 + 1] + 1 = 2^{n-2} + 5.$$

If the (single) relaxation of the "divine rule" was allowed at the time of beginning of the disc transfer process, from Theorem 2.1, this would have reduced the life-time of the world by ¾! But, if the relaxation is allowed now, the situation may be different, depending on the current situation in the temple. If the priests have not yet touched the 62nd disc, the life-time of the world would be reduced (approximately, by a factor of 1/4). However, if the priests have already formed the tower on the 62nd but have not yet touched the 63rd disc, the number of moves required to complete the process, taking into account the new Sermon of God, is $2^{63} + 1$, that is, the life-time of the world would be (very nearly) half of the original life-time. But if they have already formed (or, in the process of forming) the tower on the 63rd disc, are we safe with the new Sermon of God?

References

1. Lucas, E. (1885) Discours Prononce a la Distribution Solennelle des Prix. Faite la Mardi 4 Aout 1885. Lycee Saint Louis, Paris.
2. Ball, W.W.R. (1892) **Mathematical recreations and Essays**. MacMillan Book Co. Ltd., London.
3. Gardner, M. (1956) Mathematical Puzzles and Diversions. Penguin Book. U.K.
4. Chen, X., Tian, B. and Wang, L. (2007) Santa Claus' Towers of Hanoi. ***Graphs and Combinatorics*, 23 (Supplement),** 153 – 167.

Wordplay

Rachel Pollari

Charles Ashbacher

Written by Rachel Pollari

Copper money piece for your deliberations?

Deeds articulate more thunderously than verse.

To augment offence with grievance…

Recede to the doodling plank.

It's the superlative entity in light of divided yeast and flour product.

You've gnawed to isolation and exceeded that which you can masticate.

The expense is an appendage from the shoulder and a limb below the hip.

Exorbitant occasions harken exorbitant quotas.

Each condensed mass of water vapor floating in the atmosphere contains lustrous stuffing.

It requires a dyad to boogie in a Latin style.

A delineation renders a hundred tens' remarks.

Endure it with a granule of sodium chloride.

Written by Charles Ashbacher

Singularities are a consequence of the supreme being placing a naught as the denominator of a fraction where the numerator is the universe.

Unbounded numeration is a bottomless dwelling place with no horizontal or vertical bounds.

The Supreme Being is expressible as a Dedekind cut, unless of course defined as a whole number fraction where the denominator is one.

Mathematicians in the "School of Athens"

Charles Ashbacher

The famous painting "The School of Athens" by Rafael appears in the following figure.

This image contains representations of the following mathematical personalities.

Zeno, Epicurus, Averroes, Pythagoras, Xenophon, Socrates, Parmenides, Heraclitus, Plato, Aristotle, Diogenes, Euclid, Zoroaster and Ptolemy.

The task here is to fill in the following crossword puzzle with these 14 names.

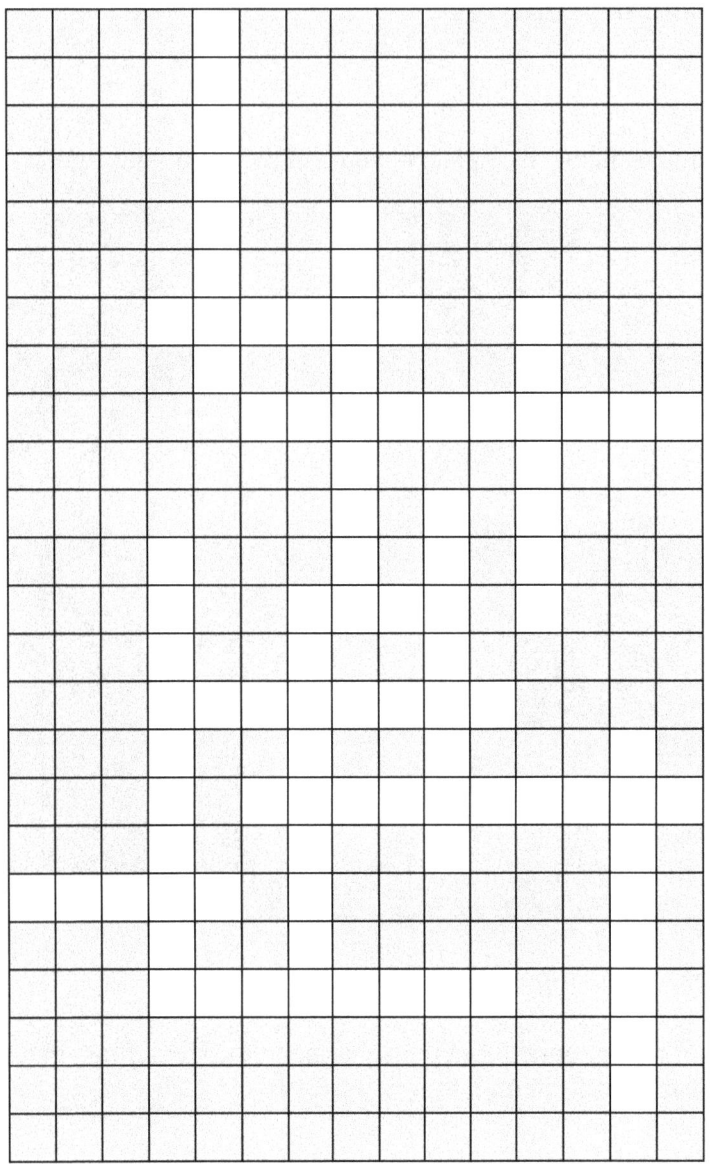

Solutions To Mathematical Kabobs

```
  COMMUTATIVE
    EQUALITY
      NEGATIVE
      IDENTITY
  DISTRIBUTIVE
     INVERSE
   ASSOCIATIVE
```

```
     ANGLE
      EUCLID
     POINT
   SEGMENT
     RECTANGLE
       TRIANGLE
   SQUARE
      Y-AXIS
```

```
R A T I O N A L
        S U R R E A L
        C O M P L E X
            B O O L E A N
          R E A L S
    N A T U R A L
I N T E G E R S
```

More Irreverence?

Charles Ashbacher

cashbacher@yahoo.com

In his article titled "Irreverence?" in volume 18(3) of **Journal of Recreational Mathematics**, Charles W. Trigg references ESPN sportscaster Chris Berman and his habit of making up nicknames for sports figures. Some of the names are Gary (Hospital) Ward, Storm (Trooper) Davis, Bruce (Eggs) Benedict and Terry (Swimming) Puhl.

Trigg then creates a list of similar nicknames using the names of mathematicians. Some of them are John (Bones) Napier, George (Canned) Salmon, Isaac (Wheel) Barrow, Isaac (Fig) Newton and Michael (Tootsie) Rolle.

In an attempt to honor yet another unusual contribution from Trigg, I created several more mathematical nicknames. They are:

Godfrey (Hale and) Hardy

Martin (Vegetable) Gardner

Richard (Family) Guy

Grigori (Knit and)Perelman

Andrew (Many) Wiles

Stephen (Tommy) Hawking

Alfred (Snow) Whitehead

Georg (Horselike) Cantor

Giuseppe (Player) Peano

Shiing-Shen (Butter) Chern

Henry J.S. (Black) Smith

Raoul (Ro) Bott

Peter David (Para) Lax

Alan (Toss and) Turing

Pierre-Simon (This must be)Laplace

Kurt (Diamond-studded) Gödel

Lewis (Christmas) Carroll

Reference

Charles W. Trigg, *Irreverence?*, **Journal of Recreational Mathematics**, 18(3), pp. 210, 1985-86.

Palindromic Numbers and Iterations of the Pseudo-Smarandache Function

Charles Ashbacher
cashbacher@yahoo.com

Abstract

For $n \geq 1$, the Pseudo-Smarandache function $Z(n)$ is the smallest integer m such that n evenly divides $1 + 2 + 3 + \ldots + m$. In this paper, some iterations of this function on palindromes that yield palindromes are demonstrated.

This paper was originally published in **Proceedings of the First International Conference on Smarandache Type Notions in Number Theory**, American Research Press, 1997. ISBN 1-879585-58-8.

In his delightful book[1], Kenichiro Kashara introduced the Pseudo-Smarandache function.

Definition: For any n ≥ 1, the value of the Pseudo-Smarandache function Z(n) is the smallest integer m such that n evenly divides 1 + 2 + 3 + . . . + m. It is well known that the sum is equivalent to

$$\frac{m(m+1)}{2}.$$

Having been defined only recently, many of the properties of this function remain to be discovered. In this short paper, we will tentatively explore the connections between Z(n) and a subset of the integers known as the palindromic numbers.

Definition: A number is said to be a palindrome if it reads the same forward and backwards. Examples of palindromes are 121, 34566543 and 1111111111.

There are some palindromic numbers n such that Z(n) is also palindromic. For example,

Z(909) = 404 and Z(2222) = 1111.

In this paper, we will not consider the trivial cases of the single digit numbers.

A simple computer program was used to search for values of n satisfying the above criteria. The range of the search was, 10 ≤ n ≤ 10000. Of the 189 palindromic values of n within that range, 37, or slightly over 19%, satisfied the criteria.

Furthermore, it is sometimes possible to repeat the function and get another palindrome.

Z(909) = 404, Z(404) = 303.

Once again, a computer program was run looking for values of n within the range 1 ≤ n ≤ 10000. Of the 37 values found in the previous test, 9 or slightly over 24%, exhibited the property of repeated palindromes.

Using the program to test for values of n such that n, Z(n), Z(Z(n)) and Z(Z(Z(n))) are all palindromic, we discovered that of the 9 found in the previous test, 2 or roughly 22% satisfy the new criteria.

Definition: Let $Z^k(n) = Z(Z(Z(...(n)...))$ where the Z function is executed k times. For notational purposes, let $Z^0(n) = n$.

Modifying the program to search for solutions for a value of n so that n and all iterations $Z^k(n)$ are palindromic for k = 1, 2, 3 and 4, we found that there were no solutions in the range $1 \leq n \leq 10000$. Given the percentages already encountered, this should not be a surprise. In fact, by expanding the search up through 100000, one solution was found.

$Z(86868) = 17271$, $Z(17271) = 2222$, $Z(2222) = 1111$, $Z(1111) = 505$.

Since $Z(505) = 100$, this is the largest such sequence in this region.

Computer searches for larger such sequences can be more efficiently carried out by using only palindromic numbers for n.

Unsolved Question: What is the largest value of m so that for some $Z^k(n)$ is a palindrome for all k = 0, 1, 2, ..., m?

Unsolved Question: Do the percentages discussed previously accurately represent the general case?

Of course, an affirmative answer to the second question would mean that there is no largest value of m.

Conjecture: There is no largest value of m such that for some n, $Z^k(n)$ is a palindrome for all k = 0, 1, 2, 3, ..., m.

There are solid arguments in support of the truth of this conjecture. Palindromes tend to be divisible by palindromic numbers, so if we take a palindromic, many of the numbers that divide it would be palindromic. Furthermore, that palindrome is often the product of two numbers, one of which is a palindrome. Numbers like the repunits, 11...11 and those with a small number of different digits, like 1001 and 505 appeared quite regularly in the computer search.

Reference

1. K. Kashihara, **Comments and Topics on Smarandache Notions and Problems**, Erhus University Press, Vail, AZ, 1996.

Divisibility and Periodicity Patterns and Palatable Number Tricks in the Jacobsthal Sequence

Jay L. Schiffman
Rowan University
schiffman@rowan.edu

Abstract

The Jacobsthal sequence is a Fibonacci-like sequence defined as follows:

$J_0 = 0$, $J_1 = 1$ and $J_n = J_{n-1} + 2 * J_{n-2}$ for $n \geq 2$.

The initial few terms of this sequence are 0, 1, 1, 3, 5, 11, 21, 43, 85,.... This paper will explore divisibility and periodicity patterns, early primes and palatable number tricks in this sequence.

In table 1, we initiate our discussion with the initial fifty terms in the Jacobsthal sequence and their prime factorizations with prime elements highlighted in bold.

Table 1

Term	Value	Prime Factorization
J_1	1	1
J_2	1	1
J_3	3	**3**
J_4	5	**5**
J_5	11	**11**
J_6	21	$3 \cdot 7$
J_7	43	**43**
J_8	85	$5 \cdot 17$
J_9	171	$3^2 \cdot 19$
J_{10}	341	$11 \cdot 31$
J_{11}	683	**683**
J_{12}	1365	$3 \cdot 5 \cdot 7 \cdot 13$
J_{13}	2731	**2731**
J_{14}	5461	$43 \cdot 127$
J_{15}	10923	$3 \cdot 11 \cdot 331$
J_{16}	21845	$5 \cdot 17 \cdot 257$
J_{17}	43691	**43691**

J_{18}	87381	$3^2 \cdot 7 \cdot 19 \cdot 73$
J_{19}	174763	**174763**
J_{20}	349525	$5^2 \cdot 11 \cdot 31 \cdot 41$
J_{21}	699051	$3 \cdot 43 \cdot 5419$
J_{22}	1398101	$23 \cdot 89 \cdot 683$
J_{23}	2796203	**2796203**
J_{24}	5592405	$3 \cdot 5 \cdot 7 \cdot 13 \cdot 17 \cdot 241$
J_{25}	11184811	$11 \cdot 251 \cdot 4051$
J_{26}	22369621	$2731 \cdot 8191$
J_{27}	44739243	$3^3 \cdot 19 \cdot 87211$
J_{28}	89478485	$5 \cdot 29 \cdot 43 \cdot 113 \cdot 127$
J_{29}	178956971	$59 \cdot 3033169$
J_{30}	357913941	$3 \cdot 7 \cdot 11 \cdot 31 \cdot 151 \cdot 331$
J_{31}	715827883	**715827883**
J_{32}	1431655765	$5 \cdot 17 \cdot 257 \cdot 65537$
J_{33}	2863311531	$3 \cdot 67 \cdot 683 \cdot 20857$
J_{34}	5726623061	$43691 \cdot 131071$
J_{35}	11453246123	$11 \cdot 43 \cdot 281 \cdot 86171$
J_{36}	22906492245	$3^2 \cdot 5 \cdot 7 \cdot 13 \cdot 19 \cdot 37 \cdot 73 \cdot 109$
J_{37}	45812984491	$1777 \cdot 25781083$
J_{38}	91625968981	$174763 \cdot 524287$

J_{39}	183251937963	$3 \cdot 2731 \cdot 22366891$
J_{40}	366503875925	$5^2 \cdot 11 \cdot 17 \cdot 31 \cdot 41 \cdot 61681$
J_{41}	733007751851	$83 \cdot 8831418697$
J_{42}	1466015503701	$3 \cdot 7^2 \cdot 43 \cdot 127 \cdot 337 \cdot 5419$
J_{43}	2932031007403	**2932031007403**
J_{44}	5864062014805	$5 \cdot 23 \cdot 89 \cdot 397 \cdot 683 \cdot 2113$
J_{45}	11728124029611	$3^2 \cdot 11 \cdot 19 \cdot 331 \cdot 18837001$
J_{46}	23456248059221	$47 \cdot 178481 \cdot 2796203$
J_{47}	46912496118443	$283 \cdot 165768537521$
J_{48}	93824992236885	$3 \cdot 5 \cdot 7 \cdot 13 \cdot 17 \cdot 97 \cdot 241 \cdot 257 \cdot 673$
J_{49}	187649984473771	$43 \cdot 4363953127297$
J_{50}	375299968947541	$11 \cdot 31 \cdot 251 \cdot 601 \cdot 1801 \cdot 4051$

An examination of the table leads one to a plethora of conjectures with regards to the Jacobsthal sequence. First note that the units digit forms the pattern 1, 1, 3, 5, 1, 1, 3, 5,... and recycles after four terms. Hence no term of the Jacobsthal sequence terminates with the digits 7 or 9 or in any even digit with the exception of the zeroth term which is 0. The sequence is rich in primes with the following terms yielding prime outputs: the third, fourth, fifth, seventh, eleventh, seventeenth, nineteenth, twenty-third, thirty-first and forty-third. As is the case with the traditional Fibonacci sequence, prime outputs arise from terms of prime index with the exception of the fourth term. The converse is not true; for the twenty-ninth term is not prime while twenty-nine is a prime number.

Table 2 provides the entry points for each of the first fifty (odd) primes in the Jacobsthal sequence.

Table 2

Prime	Entry Point in the Jacobsthal Sequence	Prime	Entry Point in the Jacobsthal Sequence
3	J_3	103	J_{102}
5	J_4	107	J_{53}
7	J_6	109	J_{36}
11	J_5	113	J_{28}
13	J_{12}	127	J_{14}
17	J_8	131	J_{65}
19	J_9	137	J_{68}
23	J_{22}	139	J_{69}
29	J_{28}	149	J_{148}
31	J_{10}	151	J_{30}
37	J_{36}	157	J_{52}
41	J_{20}	163	J_{81}
43	J_7	167	J_{166}
47	J_{46}	173	J_{172}
53	J_{52}	179	J_{89}
59	J_{29}	181	J_{180}
61	J_{60}	191	J_{190}
67	J_{33}	193	J_{96}

71	J_{70}	197	J_{196}
73	J_{18}	199	J_{198}
79	J_{78}	211	J_{105}
83	J_{41}	223	J_{74}
89	J_{22}	227	J_{113}
97	J_{48}	229	J_{76}
101	J_{100}	233	J_{58}

One might correctly conjecture that if p is prime, then p enters the Jacobsthal sequence no later than the p – 1st term or if earlier as a divisor of the p – 1st term. Using Mathematical Induction, we prove that every fifth term in the Jacobsthal sequence is divisible by eleven. Our proposition is as follows:

P(n) : For all n ≥ 1, 11 | J_{5*n}

We establish the truth of $P(1): 11 \mid J_{5 \cdot 1} \Leftrightarrow 11 \mid J_5 \Leftrightarrow 11 \mid 11$; for $11 = 11 \cdot 1$. This is the base case.

We next establish the truth of $P(k+1)$ given the assumed truth of $P(k)$.

We seek $P(k+1): 11 \mid J_{5 \cdot (k+1)}$.

$J_{5 \cdot (k+1)} = J_{5k+5} = J_{5k+4} + 2 \cdot J_{5k+3} = [J_{5k+3} + 2 \cdot J_{5k+2}] + 2 \cdot J_{5k+3} = 3 \cdot J_{5k+3} + 2 \cdot J_{5k+2} = 3 \cdot [J_{5k+2} + 2 \cdot J_{5k+1}] + 2 \cdot J_{5k+2} = 5 \cdot J_{5k+2} + 6 \cdot J_{5k+1} = 5 \cdot [J_{5k+1} + 2 \cdot J_{5k}] + 6 \cdot J_{5k+1} = 11 \cdot J_{5k+1} + 10 \cdot J_{5k}$.

Since by the induction hypothesis $11 \mid J_{5k} \Rightarrow 11 \mid 10 \cdot J_{5k}$. In addition $11 \mid 11 \Rightarrow 11 \mid 11 \cdot J_{5k+1}$. Since $11 \mid 11 \cdot J_{5k+1}$ and $11 \mid 10 \cdot J_{5k} \Rightarrow 11 \mid [11 \cdot J_{5k+1} + 10 \cdot J_{5k}] \Leftrightarrow 11 \mid J_{5k+5} \Leftrightarrow 11 \mid J_{5 \cdot (k+1)}$ which is $P(k+1)$.

By the principle of Mathematical Induction, the statement is true for all n ≥ 1.

One can also utilize modular arithmetic modulo 11 to confirm this result. Note that the sequence of remainders cycles in the pattern { 1, 2, 3, 5, 0, 10, 10, 8, 6, 0, ... }. The length of the period is ten.

Similarly the following divisibility results can be proven via mathematical induction or modular arithmetic:

a) $3 \mid J_{3n}$

b) $5 \mid J_{4n}$

c) $7 \mid J_{6n}$

d) $13 \mid J_{12n}$

e) $17 \mid J_{18n}$.

Table 3 contains the periods of the Jacobsthal numbers modulo a prime p for all of the first fifty odd primes.

Table 3

Odd Prime P	Length of Period Mod P	Odd Prime P	Length of Period Mod P
3	6	103	102
5	4	107	106
7	6	109	36
11	10	113	28
13	12	127	14
17	8	131	130
19	18	137	68
23	22	139	138
29	28	149	148
31	10	151	30
37	36	157	52

41	20	163	162
43	14	167	166
47	46	173	172
53	52	179	178
59	58	181	180
61	60	191	190
67	66	193	96
71	70	197	196
73	18	199	198
79	78	211	210
83	82	223	74
89	22	227	226
97	48	229	76
101	100	233	58

We conclude by displaying a number of palatable number tricks associated with this sequence. Consider consecutive terms in the Jacobsthal sequence and the cumulative sums. Let x and y denote two consecutive terms at any point in the sequence. The cumulative sums appear in table 4.

Table 4

Consecutive Terms in the Jacobsthal Sequence	Cumulative Sum
x	x
y	$x + y$
$y + 2 \cdot x$	$3 \cdot x + 2 \cdot y$
$3 \cdot y + 2 \cdot x$	$5 \cdot x + 5 \cdot y$
$5 \cdot y + 6 \cdot x$	$11 \cdot x + 10 \cdot y$
$11 \cdot y + 10 \cdot x$	$21 \cdot x + 21 \cdot y$
$21 \cdot y + 22 \cdot x$	$43 \cdot x + 42 \cdot y$
$43 \cdot y + 42 \cdot x$	$85 \cdot x + 85 \cdot y$
$85 \cdot y + 86 \cdot x$	$171 \cdot x + 170 \cdot y$
$171 \cdot y + 170 \cdot x$	$341 \cdot x + 341 \cdot y$
$341 \cdot y + 340 \cdot x$	$683 \cdot x + 682 \cdot y$
$683 \cdot y + 682 \cdot x$	$1365 \cdot x + 1365 \cdot y$
$1365 \cdot y + 1366 \cdot x$	$2731 \cdot x + 2730 \cdot y$
$2731 \cdot y + 2730 \cdot x$	$5461 \cdot x + 5461 \cdot y$
$5461 \cdot y + 5462 \cdot x$	$10923 \cdot x + 10922 \cdot y$
$10923 \cdot y + 10922 \cdot x$	$21845 \cdot x + 21845 \cdot y$
$21845 \cdot y + 21846 \cdot x$	$43691 \cdot x + 43690 \cdot y$
$43691 \cdot y + 43690 \cdot x$	$87381 \cdot x + 87381 \cdot y$
$87381 \cdot y + 87382 \cdot x$	$174763 \cdot x + 174762 \cdot y$
$174763 \cdot y + 174762 \cdot x$	$349525 \cdot x + 349525 \cdot y$

$349525 \cdot y + 349526 \cdot x$	$699051 \cdot x + 699050 \cdot y$
$699051 \cdot y + 699050 \cdot x$	$1398101 \cdot x + 1398101 \cdot y$
$1398101 \cdot y + 1398102 \cdot x$	$2796203 \cdot x + 2796202 \cdot y$
$2796203 \cdot y + 2796202 \cdot x$	$5592405 \cdot x + 5592405 \cdot y$
$5592405 \cdot y + 5592406 \cdot x$	$11184811 \cdot x + 11184810 \cdot y$

Let us first recall the standard Fibonacci sequence whose terms are 1, 1, 2, 3, 5, …. For example, in the Fibonacci sequence, if one adds any ten consecutive terms and divides the resulting sum by eleven, the quotient will always be the seventh term in the sequence. The following palatable number tricks are associated with the Jacobsthal sequence in much the same manner as with the standard Fibonacci sequence and we will verify several of them using simple algebra.

Trick 1: The sum of four consecutive terms is divisible by five and the quotient is the sum of the first two terms.

Trick 2: The sum of six consecutive terms is divisible by twenty-one and the quotient is the sum of the first two terms.

Trick 3: The sum of six consecutive terms is divisible by seven and the quotient is the sum of the first and fourth terms.

Trick 4: The sum of eight consecutive terms is divisible by eighty-five and the quotient is the sum of the first two terms.

Trick 5: The sum of eight consecutive terms is divisible by seventeen and the quotient is the sum of the first four terms.

Trick 6: The sum of eight consecutive terms is divisible by five and the quotient is the sum of the first, second, fifth and sixth terms.

Trick 7: The sum of twelve consecutive terms is divisible by one thousand three hundred sixty-five and the quotient is the sum of the first two terms.

Trick 8: The sum of twelve consecutive terms is divisible by five and the quotient is the sum of the first, second, fifth, six, ninth and tenth terms.

Trick 9: The sum of twelve consecutive terms is divisible by seven and the quotient is the sum of the first, fourth, seventh and tenth terms.

Trick 10: The sum of twelve consecutive terms is divisible by thirteen and the quotient is the sum of the first, second, fourth, fifth, sixth and ninth terms.

Trick 11: The sum of twelve consecutive terms is divisible by sixty-five and the quotient is the sum of the first six terms.

Trick 12: The sum of twelve consecutive terms is divisible by twenty-one and the quotient is the sum of the first, second, seventh and eighth terms.

Trick 13: The sum of twelve consecutive terms is divisible by thirty-nine and the quotient is the sum of the first, fourth, sixth and seventh terms.

Trick 14: The sum of twelve consecutive terms is divisible by ninety-one and the quotient is the sum of the first, third, fourth and sixth terms.

Trick 15: The sum of twelve consecutive terms is divisible by two hundred seventy-three and the quotient is the sum of the first four terms.

Trick 16: The sum of twelve consecutive terms is divisible by four hundred fifty-five and the quotient is the sum of the first and fourth terms.

Trick 17: The sum of fourteen consecutive terms is divisible by five thousand four hundred sixty-one and the quotient is the sum of the first two terms.

Trick 18: The sum of fourteen consecutive terms is divisible by one hundred twenty-seven and the quotient is the sum of the first and eighth terms.

Trick 19: The sum of sixteen consecutive terms is divisible by five and the quotient is the sum of the first, second, fifth, sixth, ninth, tenth, thirteenth and fourteenth terms.

Trick 20: The sum of sixteen consecutive terms is divisible by twenty one thousand eight hundred forty-five and the quotient is the sum of the first two terms.

Trick 21: The sum of sixteen consecutive terms is divisible by four thousand three hundred sixty- nine and the quotient is the sum of the first four terms.

Trick 22: The sum of sixteen consecutive terms is divisible by one thousand two hundred eighty- five and the quotient is the sum of the first, second, fifth and sixth terms.

Trick 23: The sum of sixteen consecutive terms is divisible by eighty-five and the quotient is the sum of the first, second, ninth and tenth terms.

Trick 24: The sum of sixteen consecutive terms is divisible by two hundred fifty-seven and the quotient is the sum of the first eight terms.

We will prove three of these number tricks, numbers 3, 8 and 14.

Trick 3: The sum of six consecutive terms is divisible by seven and the quotient is the sum of the first and fourth terms.

Proof: Let x and y be the next two terms at any point in the sequence. Then

x + y + y + 2 * x + 3 * y + 2 * x + 5 * y + 6 * x + 11 * y + 10 * x = 21 * x + 21 * y.

This is easily seen to be divisible by 7 with the quotient 3 * x + 3 * y = x + 3 * y + 2 * x.

Trick 8: The sum of twelve consecutive terms is divisible by five and the quotient is the sum of the first, second, fifth, six, ninth and tenth terms.

Proof: Let x and y be the next two terms at any point in the sequence. Then

x + y + y + 2 * x + 3 * y + 2 * x + 5 * y + 6 * x+ 11 * y + 10 * x + 21 * y + 22 * x + 43 * y +

42 * x + 85 * y + 86 * x + 171 * y * 170 * x + 341 * y + 342 * x + 683 * y + 682 * x =

1365 * x + 1365 * y.

This is obviously evenly divisible by 5 and the quotient is 273 * x + 273 * y. This is equal to

x + y + 5 * y + 6 * x + 11 * y + 10 * x + 85 * y + 86 * x + 171 * y + 170 * x.

Trick 14: The sum of twelve consecutive terms is divisible by ninety-one and the quotient is the sum of the first, third, fourth and sixth terms.

Proof: Let x and y be the next two terms at any point in the sequence. Then

x + y + y + 2 * x + 3 * y + 2 * x + 5 * y + 6 * x + 11 * y + 10 * x + 21 * y + 22 * x + 43 * y +

42 * x + 85 * y + 86 * x + 171 * y + 170 * x + 341 * y + 342 * x + 683 * y + 682 * x =

= 1365 * x + 1365 * y.

This is easily seen to be divisible by 91 and the quotient is 15 * x + 15 * y. This is equal to

x + y + 2 * x + 3 * y + 2 * x + 11 * y + 10 * x.

Conclusion

This paper served to explore palatable number tricks as well as divisibility and periodicity patterns and prime outputs associated with the Jacobsthal sequence. Readers are encouraged to explore a world of possibilities with this fun recreational mathematics topic.

References

1. Mathematica v. 9.0, *Wolfram Research, Inc.,* Champaign, IL. (2016)
2. *MathWorld-A Wolfram Resource,* Wolfram Research, Inc, Champaign, IL. (2016)
3. The On-Line Encyclopedia of Integer Sequences (www. oeis.org) (2016); sequence A001045.

Abstracts of the papers in "Journal of Recreational Mathematics" Volume 1, Number 1, 1968

Charles Ashbacher
cashbacher@yahoo.com

Given the elapsed time since these papers appeared and that there were no abstracts with the originals, all of the items in this list were written by Charles Ashbacher.

"Magic Designs," by Robert B. Ely, III

Abstract
The definition of a magic design is as follows:

A design with N parts is said to be magic, if those parts can be labeled with the numbers 1 to N so that the labels of each of a number of identical sub-designs give the same constant total.

A magic square, which is an m x m grid where each row and column has the same sum is the most widely known magic design.
In this paper, additional magic designs, such as a triangle, cube, a grid made of hexagons, a triangular grid and the faces of a dodecahedron are analyzed.

"Counting Planar Maps," by W. T. Tutte

Abstract
The paper opens with the examination of a standard die as an object having six faces. Such an object is called a hexahedra. There are seven convex objects having exactly six faces and three that are concave. Schlegel diagrams are used to convert the polyhedra into planar graphs for further analysis. Planar diagrams that can be defined by 3-connected graphs are called c-nets. A table of the number of c-nets for each number of edges for n from 4 through 25 is given along with the explicit formula used to compute the numbers.

"Infinite Geometry," by Donald L. Vanderpool

Abstract
While Euclidean geometry is performed on a plane infinite in two directions, triangles are described as being finite. In this paper, the lines that form the triangles are geodesic's in Einstein's 4-dimensional space. This leads to three-sided figures where the sums of the angles are vastly different from the standard 180 degrees.

"A Recurrent Operation Leading to a Number Trick," by Charles W. Trigg

Abstract
For each two-digit number S, the digits are written in reverse order. This is added to the sum of the digits of S and then the sum is reduced modulo 100. The computations are then repeated.

Under this operation, the fifty-two odd two-digit numbers form a series in which the last 6 numbers form a closed loop.

"Alphametics," by J. A. H. Hunter

Abstract
This paper re-introduces the letter-arithmetic puzzle of alphametic and gives a brief description of the logic one uses to solve them. The example problem is a division alphametic.

"The Witch of Agnesi," by Harold D. Larsen

Abstract
The Witch of Agnesi is a curve defined by the equations:

$$y(a^2 + x^2) = a^3 \text{ or } x = a(\cot \theta), y = a(\sin \theta)^2.$$

The history and some of the major properties of this curve are examined.

"Curiosa for 1968," by Charles W. Trigg

Abstract
This paper gives several ways that 1968 can be written using conventional math symbols and the nine decimal digits.

"Pentomino Farms," by Fr. Victor Feser, OSB

Abstract
The 12 pentominoes are formed from five squares being connected edge-to-edge. A pentomino farm is a rectangular shape made by the set of pentominoes that has enclosed open space(s). Several of the most basic are presented.

"Squares with 9 and 10 Distinct Digits," by T. Charles Jones

Abstract
In this paper, two computer generated tables are given. The first contains all squares with nine distinct digits and the second all squares with ten distinct digits.

Smarandache Bisymmetric Geometric Determinant Sequence

A.A.K. Majumdar

APU, 1–1 Jumonjibaru, Beppu-shi 875–8577, Oita-ken, Japan
majumdar@apu.ac.jp/aakmajumdar@gmail.com

Abstract
In this paper, the concept of Smarandache bisymmetric geometric determinant sequence has been introduced. An explicit form of the nth term is given.

Key Words Smarandache bisymmetric geometric determinant sequence, n^{th} term of the sequence.

Introduction

Murthy[1] introduced the concept of the Smarandache cyclic arithmetic determinant sequence and the Smarandache bisymmetric arithmetic determinant sequence. The nth terms of these sequences are given in Majumdar[2] and independently by Maohua Le [4] in a slightly different form. In a recent paper, Bueno[3] extended the Smarandache cyclic arithmetic determinant sequence to the Smarandache cyclic geometric sequences of right circulant and left circulant forms, and derived formulas for the nth terms of the sequences. This paper extends the Smarandache bisymmetric arithmetic sequence to the geometric case.

Definition: The Smarandache bisymmetric geometric determinant sequence, denoted by $\{SBGDS(n)\}$, is

$$\left\{ |1|, \begin{vmatrix} 1 & r \\ r & 1 \end{vmatrix}, \begin{vmatrix} 1 & r & r^2 \\ r & r^2 & r \\ r^2 & r & 1 \end{vmatrix}, \ldots \right\}.$$

The first few terms of this sequence are

$$1,\ 1 - r^2,\ -r^2(r^2 - 1)^2,\ r^6(r^2 - 1)^3,\ r^{12}(r^2 - 1)^4,\ -r^{20}(r^2 - 1)^5,\ \ldots.$$

In this paper, we derive the explicit form of the nth term of the sequence and this is given in Section 3. Some preliminary results, that are necessary in the derivation of the expressions of the nth terms of the sequences are given in section 2.

2. Some Preliminary Results

In this section, we derive some results that will be needed later in proving the main results of this paper in section 3. We start with the following result.

Lemma 2.1 Let $D \equiv |d_{ij}|$ be the determinant of order n (≥ 2) with

$$d_{ij} = \begin{cases} 1, & \text{if } j = n - i + 1 \\ 0, & \text{otherwise} \end{cases} \quad \text{for } 1 \leq i, j \leq n.$$

Then,

$$D \equiv \begin{vmatrix} 0 & 0 & \cdots & 0 & 0 & 1 \\ 0 & 0 & \cdots & 0 & 1 & 0 \\ 0 & 0 & \cdots & 1 & 0 & 0 \\ \vdots & \vdots & \cdots & \vdots & \vdots & \vdots \\ 0 & 0 & \cdots & 0 & 0 & 0 \\ 1 & 0 & 0 & 0 & 0 & 0 \end{vmatrix} = (-1)^{\left[\frac{n}{2}\right]}.$$

Proof: First, let n be odd, say, n = 2m + 1 (for m $\geq$ 1). Then, using the m column operations $C_j \leftrightarrow C_{2m-j+2}$ ($1 \leq j \leq m$), we get

$$D = (-1)^m \begin{vmatrix} 1 & 0 & \cdots & 0 & 0 & 0 \\ 0 & 1 & \cdots & 0 & 0 & 0 \\ 0 & 0 & \cdots & 1 & 0 & 0 \\ \vdots & \vdots & \cdots & \vdots & \vdots & \vdots \\ 0 & 0 & \cdots & 0 & 1 & 0 \\ 0 & 0 & 0 & 0 & 0 & 1 \end{vmatrix} = (-1)^m.$$

Now, let n = 2m (m $\geq$ 1). Then, using the column operations $C_j \leftrightarrow C_{2m-j+1}$ ($1 \leq j \leq m$), we get $D = (-1)^m$. Since, in either case, $m = \left[\frac{n}{2}\right]$, and the lemma is established.

Lemma 2.2 Let $D \equiv |d_{ij}|$ be the determinant of order 2m + 2 (m $\geq$ 1) with

$$d_{ij} = \begin{cases} 1, & \text{if } j = 2m - i + 3 \\ r, & \text{if } i = m+1, j = m+2, \text{ or } i = m+2, j = m+1 \\ 0, & \text{otherwise} \end{cases}.$$

Then,

$$D = \begin{vmatrix} 0 & 0 & 0 & \cdots & 0 & 0 & \cdots & 0 & 0 & 1 \\ 0 & 0 & 0 & \cdots & 0 & 0 & \cdots & 0 & 1 & 0 \\ \vdots & \vdots & \vdots & & \vdots & \vdots & & \vdots & \vdots & \vdots \\ 0 & 0 & 0 & \cdots & 1 & r & \cdots & 0 & 0 & 0 \\ 0 & 0 & 0 & \cdots & r & 1 & \cdots & 0 & 0 & 0 \\ \vdots & \vdots & \vdots & & \vdots & \vdots & & \vdots & \vdots & \vdots \\ 0 & 1 & 0 & \cdots & 0 & 0 & \cdots & 0 & 0 & 0 \\ 1 & 0 & 0 & \cdots & 0 & 0 & \cdots & 0 & 0 & 0 \end{vmatrix} = (-1)^{m+1}(r^2 - 1).$$

3. Main result

We now give the main result of this paper.

Theorem 3.1. The nth term, SBGDS(n), of the Smarandache bisymmetric geometric determinant sequence is given by

$$SBGDS(n) = \begin{vmatrix} 1 & r & r^2 & \cdots & r^{n-1} & r^{n-2} & r^{n-1} \\ r & r^2 & r^3 & \cdots & r^{n-2} & r^{n-1} & r^{n-2} \\ r^2 & r^3 & r^4 & \cdots & r^{n-1} & r^{n-2} & r^{n-3} \\ \vdots & \vdots & \vdots & & \vdots & \vdots & \vdots \\ r^{n-3} & r^{n-2} & r^{n-1} & \cdots & r^4 & r^3 & r^2 \\ r^{n-2} & r^{n-1} & r^{n-2} & \cdots & r^3 & r^2 & r \\ r^{n-1} & r^{n-2} & r^{n-3} & \cdots & r^2 & r & 1 \end{vmatrix}$$

$$= (-1)^{\left\lfloor \frac{n}{2} \right\rfloor} r^{(n-1)(n-2)} (1 - r^2)^{n-1}.$$

Proof: To prove the theorem, we need to consider the two cases separately below.

Case 1. When n is odd, say, n = 2m + 1 (for some m ≥ 1). In this case,

SBGDS(2m + 1)

$$= \begin{vmatrix} 1 & r & r^2 & \cdots & r^{m-1} & r^m & r^{m+1} & \cdots & r^{2m-2} & r^{2m-1} & r^{2m} \\ r & r^2 & r^3 & \cdots & r^m & r^{m+1} & r^{m+2} & \cdots & r^{2m-1} & r^{2m} & r^{2m-1} \\ r^2 & r^3 & r^4 & \cdots & r^{m+1} & r^{m+2} & r^{m+3} & \cdots & r^{2m} & r^{2m-1} & r^{2m-2} \\ \vdots & \vdots & \vdots & & \vdots & \vdots & \vdots & & \vdots & \vdots & \vdots \\ r^{m-1} & r^m & r^{m+1} & \cdots & r^{2m-2} & r^{2m-1} & r^{2m} & \cdots & r^{m+3} & r^{m+2} & r^{m+1} \\ r^m & r^{m+1} & r^{m+2} & \cdots & r^{2m-1} & r^{2m} & r^{2m-1} & \cdots & r^{m+2} & r^{m+1} & r^m \\ r^{m+1} & r^{m+2} & r^{m+3} & \cdots & r^{2m} & r^{2m-1} & r^{2m-2} & \cdots & r^{m+1} & r^m & r^{m-1} \\ \vdots & \vdots & \vdots & & \vdots & \vdots & \vdots & & \vdots & \vdots & \vdots \\ r^{2m-2} & r^{2m-1} & r^{2m} & \cdots & r^{m+3} & r^{m+2} & r^{m+1} & \cdots & r^4 & r^3 & r^2 \\ r^{2m-1} & r^{2m} & r^{2m-1} & \cdots & r^{m+2} & r^{m+1} & r^m & \cdots & r^3 & r^2 & r \\ r^{2m} & r^{2m-1} & r^{2m-2} & \cdots & r^{m+1} & r^m & r^{m-1} & \cdots & r^2 & r & 1 \end{vmatrix}$$

$$= r^{m^2} \begin{vmatrix} 1 & 1 & 1 & \cdots & 1 & 1 & r^2 & \cdots & r^{2m-4} & r^{2m-2} & r^{2m} \\ r & r & r & \cdots & r & r & r^3 & \cdots & r^{2m-3} & r^{2m-1} & r^{2m-1} \\ r^2 & r^2 & r^2 & \cdots & r^2 & r^2 & r^4 & \cdots & r^{2m-2} & r^{2m-2} & r^{2m-2} \\ \vdots & \vdots & \vdots & & \vdots & \vdots & \vdots & & \vdots & \vdots & \vdots \\ r^{m-1} & r^{m-1} & r^{m-1} & \cdots & r^{m-1} & r^{m-1} & r^{m+1} & \cdots & r^{m+1} & r^{m+1} & r^{m+1} \\ r^m & r^m & r^m & \cdots & r^m & r^m & r^m & \cdots & r^m & r^m & r^m \\ r^{m+1} & r^{m+1} & r^{m+1} & \cdots & r^{m+1} & r^{m-1} & r^{m-1} & \cdots & r^{m-1} & r^{m-1} & r^{m-1} \\ \vdots & \vdots & \vdots & & \vdots & \vdots & \vdots & & \vdots & \vdots & \vdots \\ r^{2m-2} & r^{2m-2} & r^{2m-2} & \cdots & r^4 & r^2 & r^2 & \cdots & r^2 & r^2 & r^2 \\ r^{2m-1} & r^{2m-1} & r^{2m-3} & \cdots & r^3 & r & r & \cdots & r & r & r \\ r^{2m} & r^{2m-2} & r^{2m-4} & \cdots & r^2 & 1 & 1 & \cdots & 1 & 1 & 1 \end{vmatrix}$$

where the 2nd determinant is obtained from the 1st one by taking out the common factor r from the 2nd and (2m)th columns, r^2 from the 3rd and $(2m-1)$st columns, ..., r^{m-1} from the mth and $(m+2)$nd columns, and r^m from the $(m+1)$st column, so that the exponent of r is

$$2(1 + 2 + \ldots + (m-1)) + m = m^2.$$

Now, taking out the common factor r from the 2nd and (2m)th rows, r^2 from the 3rd and $(2m-1)$st rows, ..., r^{m-1} from the mth and $(m+2)$nd rows, and r^m from the $(m+1)$st row, we get

SBGDS(2m + 1)

$$= r^{2m^2} \begin{vmatrix} 1 & 1 & 1 & \cdots & 1 & 1 & r^2 & \cdots & r^{2m-4} & r^{2m-2} & r^{2m} \\ 1 & 1 & 1 & \cdots & 1 & 1 & r^2 & \cdots & r^{2m-4} & r^{2m-2} & r^{2m-2} \\ 1 & 1 & 1 & \cdots & 1 & 1 & r^2 & \cdots & r^{2m-4} & r^{2m-4} & r^{2m-4} \\ \vdots & \vdots & \vdots & & \vdots & \vdots & \vdots & & \vdots & \vdots & \vdots \\ 1 & 1 & 1 & \cdots & 1 & 1 & r^2 & \cdots & r^2 & r^2 & r^2 \\ 1 & 1 & 1 & \cdots & 1 & 1 & 1 & \cdots & 1 & 1 & 1 \\ r^2 & r^2 & r^2 & \cdots & r^2 & 1 & 1 & \cdots & 1 & 1 & 1 \\ \vdots & \vdots & \vdots & & \vdots & \vdots & \vdots & & \vdots & \vdots & \vdots \\ r^{2m-4} & r^{2m-4} & r^{2m-4} & \cdots & r^2 & 1 & 1 & \cdots & 1 & 1 & r^2 \\ r^{2m-2} & r^{2m-2} & r^{2m-4} & \cdots & r^2 & 1 & 1 & \cdots & 1 & 1 & r \\ r^{2m} & r^{2m-2} & r^{2m-4} & \cdots & r^2 & 1 & 1 & \cdots & 1 & 1 & 1 \end{vmatrix}$$

We now perform the m row operations $R_i \to R_i - R_{i+1}$ ($1 \leq i \leq m$) as well as the m row operations as well as the m row operations $R_{m+i+1} \to R_{m+i+1} - R_{m+i}$ ($1 \leq i \leq m$) to get

SBGDS(2m + 1)

$$= r^{2m^2} \begin{vmatrix} 0 & 0 & \cdots & 0 & 0 & 0 & \cdots & 0 & r^{2m-2}(r^2-1) \\ 0 & 0 & \cdots & 0 & 0 & 0 & \cdots & r^{2m-4}(r^2-1) & r^{2m-4}(r^2-1) \\ 0 & 0 & \cdots & 0 & 0 & 0 & \cdots & r^{2m-6}(r^2-1) & r^{2m-6}(r^2-1) \\ \vdots & \vdots & & \vdots & \vdots & \vdots & & \vdots & \vdots \\ 0 & 0 & \cdots & 0 & 0 & r^2-1 & \cdots & r^2-1 & r^2-1 \\ 1 & 1 & \cdots & 1 & 1 & 1 & \cdots & 1 & 1 \\ r^2-1 & r^2-1 & \cdots & r^2-1 & 0 & 0 & \cdots & 0 & 0 \\ \vdots & \vdots & & \vdots & \vdots & \vdots & & \vdots & \vdots \\ r^{2m-6}(r^2-1) & r^{2m-6}(r^2-1) & \cdots & 0 & 0 & 0 & \cdots & 0 & 0 \\ r^{2m-4}(r^2-1) & r^{2m-4}(r^2-1) & \cdots & 0 & 0 & 0 & \cdots & 0 & 0 \\ r^{2m-2}(r^2-1) & 0 & \cdots & 0 & 0 & 0 & \cdots & 0 & 0 \end{vmatrix}$$

From the 1st and (2m + 1)st rows, we take out the common factor $r^{2m-2}(r^2-1)$ from the 2nd and (2m)th rows, we take out the common factor $r^{2m-4}(r^2-1)$, …, and from the mth and (m + 2)nd rows, we take out the common factor (r^2-1), so that, the exponent of r is

2[2m – 2) + (2m – 4) + . . . + 2] = 2m(m - 1).

Therefore,

SBGDS(2m + 1)

$$= r^{2m^2 + 2m(m-1)}(r^2-1)^{2m} \begin{vmatrix} 0 & 0 & 0 & \cdots & 0 & 0 & 0 & \cdots & 0 & 0 & 1 \\ 0 & 0 & 0 & \cdots & 0 & 0 & 0 & \vdots & 0 & 1 & 1 \\ 0 & 0 & 0 & \cdots & 0 & 0 & 0 & \cdots & 1 & 1 & 1 \\ \vdots & \vdots & \vdots & & \vdots & \vdots & \vdots & & \vdots & \vdots & \vdots \\ 0 & 0 & 0 & \cdots & 0 & 0 & 1 & \cdots & 1 & 1 & 1 \\ 1 & 1 & 1 & \cdots & 1 & 1 & 1 & \cdots & 1 & 1 & 1 \\ 1 & 1 & 1 & \cdots & 1 & 0 & 0 & \cdots & 0 & 0 & 0 \\ \vdots & \vdots & \vdots & & \vdots & \vdots & \vdots & & \vdots & \vdots & \vdots \\ 1 & 1 & 1 & \cdots & 0 & 0 & 0 & \cdots & 0 & 0 & 0 \\ 1 & 1 & 0 & \cdots & 0 & 0 & 0 & \cdots & 0 & 0 & 0 \\ 1 & 0 & 0 & \cdots & 0 & 0 & 0 & \cdots & 0 & 0 & 0 \end{vmatrix}$$

Finally, we perform the column operations $C_i \to C_i - C_{i+1}$ ($1 \leq i \leq m$) as well as the column operations $C_{m+i+1} \to C_{m+i+1} - C_{m+i}$ ($1 \leq i \leq m$) to get

SBGDS(2m + 1)

$$= r^{2m^2+2m(m-1)} (r^2-1)^{2m} \begin{vmatrix} 0 & 0 & 0 & \cdots & 0 & 0 & 0 & \cdots & 0 & 0 & 1 \\ 0 & 0 & 0 & \cdots & 0 & 0 & 0 & \cdots & 0 & 1 & 0 \\ 0 & 0 & 0 & \cdots & 0 & 0 & 0 & \cdots & 1 & 0 & 0 \\ \vdots & \vdots & \vdots & & \vdots & \vdots & \vdots & & \vdots & \vdots & \vdots \\ 0 & 0 & 0 & \cdots & 0 & 0 & 1 & \cdots & 0 & 0 & 0 \\ 0 & 0 & 0 & \cdots & 0 & 1 & 0 & \cdots & 0 & 0 & 0 \\ 0 & 0 & 0 & \cdots & 1 & 0 & 0 & \cdots & 0 & 0 & 0 \\ \vdots & \vdots & \vdots & & \vdots & \vdots & \vdots & & \vdots & \vdots & \vdots \\ 0 & 0 & 1 & \cdots & 0 & 0 & 0 & \cdots & 0 & 0 & 0 \\ 0 & 1 & 0 & \cdots & 0 & 0 & 0 & \cdots & 0 & 0 & 0 \\ 1 & 0 & 0 & \cdots & 0 & 0 & 0 & \cdots & 0 & 0 & 0 \end{vmatrix}.$$

Now, applying Lemma 2.1, we have

SBGDS(2m + 1) = $(-1)^m r^{2m(2m-1)} (r^2-1)^{2m}$.

Case 2. When n is even, say, n = 2m + 2 (for some m ≥ 1). In this case,

SBGDS(2m + 2)

$$= \begin{vmatrix} 1 & r & r^2 & \cdots & r^m & r^{m+1} & \cdots & r^{2m-1} & r^{2m} & r^{2m+1} \\ r & r^2 & r^3 & \cdots & r^{m+1} & r^{m+2} & \cdots & r^{2m} & r^{2m+1} & r^{2m} \\ r^2 & r^3 & r^4 & \cdots & r^{m+2} & r^{m+3} & \cdots & r^{2m+1} & r^{2m} & r^{2m-1} \\ \vdots & \vdots & \vdots & & \vdots & \vdots & & \vdots & \vdots & \vdots \\ r^m & r^{m+1} & r^{m+2} & \cdots & r^{2m} & r^{2m+1} & \cdots & r^{m+3} & r^{m+2} & r^{m+1} \\ r^{m+1} & r^{m+2} & r^{m+3} & \cdots & r^{2m-1} & r^{2m} & \cdots & r^{m+2} & r^{m+1} & r^m \\ \vdots & \vdots & \vdots & & \vdots & \vdots & & \vdots & \vdots & \vdots \\ r^{2m-1} & r^{2m} & r^{2m+1} & \cdots & r^{m+3} & r^{m+2} & \cdots & r^4 & r^3 & r^2 \\ r^{2m} & r^{2m+1} & r^{2m} & \cdots & r^{m+2} & r^{m+1} & \cdots & r^3 & r^2 & r \\ r^{2m+1} & r^{2m} & r^{2m-1} & \cdots & r^{m+1} & r^m & \cdots & r^2 & r & 1 \end{vmatrix}.$$

Taking out the common factor r from the 2nd and (2m + 1)st columns, r^2 from the 3rd and (2m)th columns, …, r^m from the (m + 1)st and (m + 2)nd columns, and then taking out the common factor r from the 2nd and (2m + 1)st rows, r^2 from the 3rd and (2m)th rows, …, r^m from the (m + 1)st and (m + 2)nd rows, we get

SBGDS(2m + 2)

$$= r^{m(m+1)} \begin{vmatrix} 1 & 1 & 1 & \cdots & 1 & r & \cdots & r^{2m-3} & r^{2m-1} & r^{2m+1} \\ r & r & r & \cdots & r & r^2 & \cdots & r^{2m-2} & r^{2m} & r^{2m} \\ r^2 & r^2 & r^2 & \cdots & r^2 & r^3 & \cdots & r^{2m-1} & r^{2m-1} & r^{2m-1} \\ \vdots & \vdots & \vdots & & \vdots & \vdots & & \vdots & \vdots & \vdots \\ r^m & r^m & r^m & \cdots & r^m & r^{m+1} & \cdots & r^{m+1} & r^{m+1} & r^{m+1} \\ r^{m+1} & r^{m+1} & r^{m+1} & \cdots & r^{m+1} & r^m & \cdots & r^m & r^m & r^m \\ \vdots & \vdots & \vdots & & \vdots & \vdots & & \vdots & \vdots & \vdots \\ r^{2m-1} & r^{2m-1} & r^{2m-1} & \cdots & r^3 & r^2 & \cdots & r^2 & r^2 & r^2 \\ r^{2m} & r^{2m} & r^{2m-2} & \cdots & r^2 & r & \cdots & r & r & r \\ r^{2m+1} & r^{2m-1} & r^{2m-3} & \cdots & r & 1 & \cdots & 1 & 1 & 1 \end{vmatrix}$$

$$= r^{2m(m+1)} \begin{vmatrix} 1 & 1 & 1 & \cdots & 1 & r & \cdots & r^{2m-3} & r^{2m-1} & r^{2m+1} \\ 1 & 1 & 1 & \cdots & 1 & r & \cdots & r^{2m-3} & r^{2m-1} & r^{2m-1} \\ 1 & 1 & 1 & \cdots & 1 & r & \cdots & r^{2m-3} & r^{2m-3} & r^{2m-3} \\ \vdots & \vdots & \vdots & & \vdots & \vdots & & \vdots & \vdots & \vdots \\ 1 & 1 & 1 & \cdots & 1 & r & \cdots & r & r & r \\ r & r & r & \cdots & r & 1 & \cdots & 1 & 1 & 1 \\ \vdots & \vdots & \vdots & & \vdots & \vdots & & \vdots & \vdots & \vdots \\ r^{2m-3} & r^{2m-3} & r^{2m-3} & \cdots & r & 1 & \cdots & 1 & 1 & 1 \\ r^{2m-1} & r^{2m-1} & r^{2m-3} & \cdots & r & 1 & \cdots & 1 & 1 & 1 \\ r^{2m+1} & r^{2m-1} & r^{2m-3} & \cdots & r & 1 & \cdots & 1 & 1 & 1 \end{vmatrix}$$

Now, performing the m row operations $R_i \to R_i - R_{i+1}$, $1 \leq i \leq m$) as well as the m row operations
$R_{m+i+1} \to R_{m+i+1} - R_{m+i}$, ($1 \leq i \leq m$), we get

SBGDS(2m + 2)

$$= r^{2m(m+1)} \begin{vmatrix} 0 & 0 & 0 & \cdots & 0 & 0 & 0 & r^{2m-1}(r^2-1) \\ 0 & 0 & 0 & \cdots & 0 & 0 & r^{2m-3}(r^2-1) & r^{2m-3}(r^2-1) \\ \vdots & \vdots & \vdots & & \vdots & \vdots & \vdots & \vdots \\ 1 & 1 & 1 & 1 & r & r & & r \\ r & r & r & r & 1 & 1 & & 1 \\ \vdots & \vdots & \vdots & \vdots & \vdots & \vdots & & \vdots \\ r^{2m-3}(r^2-1) & r^{2m-3}(r^2-1) & 0 & 0 & 0 & 0 & & 0 \\ r^{2m-1}(r^2-1) & 0 & 0 & 0 & 0 & 0 & & 0 \end{vmatrix}$$

In the above determinant, we take out the common factor $r^{2m-1}(r^2-1)$ from the 1st and (2m + 2)nd rows, $r^{2m-3}(r^2-1)$ from the 2nd and (2m + 1)st rows, …, and $r(r^2-1)$ from the mth and (m + 3)rd rows, so that, the exponent of r is

$2[2m-1) + (2m-3) + \ldots + 1] = 2m^2$.

Therefore,
SBGDS(2m + 2)

$$= r^{2m(2m+1)}(r^2-1)^{2m} \begin{vmatrix} 0 & 0 & 0 & \cdots & 0 & 0 & \cdots & 0 & 0 & 1 \\ 0 & 0 & 0 & \cdots & 0 & 0 & \cdots & 0 & 1 & 1 \\ 0 & 0 & 0 & \cdots & 0 & 0 & \cdots & 1 & 1 & 1 \\ \vdots & \vdots & \vdots & & \vdots & \vdots & & \vdots & \vdots & \vdots \\ 1 & 1 & 1 & \cdots & 1 & r & \cdots & r & r & r \\ r & r & r & \cdots & r & 1 & \cdots & 1 & 1 & 1 \\ \vdots & \vdots & \vdots & & \vdots & \vdots & & \vdots & \vdots & \vdots \\ 1 & 1 & 1 & \cdots & 0 & 0 & \cdots & 0 & 0 & 0 \\ 1 & 1 & 0 & \cdots & 0 & 0 & \cdots & 0 & 0 & 0 \\ 1 & 0 & 0 & \cdots & 0 & 0 & \cdots & 0 & 0 & 0 \end{vmatrix}.$$

Finally, we perform the m column operations $C_i \rightarrow C_i - C_{i+1}$, $(1 \leq i \leq m)$ and the m column operations $C_{m+i+2} \rightarrow C_{m+i+2} - C_{m+i+1}$, $(1 \leq i \leq m)$ to get

SBGDS(2m + 2)

$$= r^{2m(2m+1)}(r^2-1)^{2m} \begin{vmatrix} 0 & 0 & 0 & \cdots & 0 & 0 & \cdots & 0 & 0 & 1 \\ 0 & 0 & 0 & \cdots & 0 & 0 & \cdots & 0 & 1 & 0 \\ 0 & 0 & 0 & \cdots & 0 & 0 & \cdots & 1 & 0 & 0 \\ \vdots & \vdots & \vdots & & \vdots & \vdots & & \vdots & \vdots & \vdots \\ 0 & 0 & 0 & \cdots & 1 & r & \cdots & 0 & 0 & 0 \\ 0 & 0 & 0 & \cdots & r & 1 & \cdots & 0 & 0 & 0 \\ \vdots & \vdots & \vdots & & \vdots & \vdots & & \vdots & \vdots & \vdots \\ 0 & 0 & 1 & \cdots & 0 & 0 & \cdots & 0 & 0 & 0 \\ 0 & 1 & 0 & \cdots & 0 & 0 & \cdots & 0 & 0 & 0 \\ 1 & 0 & 0 & \cdots & 0 & 0 & \cdots & 0 & 0 & 0 \end{vmatrix}$$

$= (-1)^{m+1} r^{2m(2m+1)} (r^2 - 1)^{2m+1}$,

where the last result follows from lemma 2.2.

All these complete the proof of the theorem.

References

1. Amarnath Murthy, "Smarandache Determinant Sequences," **Smarandache Notions Journal**, **12** (2001), 275 – 278.

2. A.A.K. Majumdar, "On Some Smarandache Determinant Sequences," **Scientia Magna, 4** (2008), 89 – 95.

3. A.C.F. Bueno, "Smarandache Cyclic Geometric Determinant Sequences," **Scientia Magna, 8** (2012), 88 – 91.

4. Maohua Le, "Two Classes of Smarandache Determinants," **Scientia Magna, 2(1)** (2006), 20 – 25.

Solutions to Elementary Problems

The solution to the problem by Szilagyi is

ABCDEFGHIJK = 1234567890

```
      1 2 6
    8 5 9 1
    8 6 7 6
      8 3 9
    7 3 0 9
   ──────────
  2 5 5 4 1
```

```
           1   2  7
        5  6   1 11
           1  11 10
       11 10   9  7
               4  3
      ─────────────
     1  5  8   6  2
```

Solution to the logic puzzle

One drinks all three, another only wine and the third only beer.

The NFL Draft, 2002-2014: Winners, Losers, and a New Draft Trade Value Chart

Paul M. Sommers

Middlebury College

psommers@middlebury.edu

Abstract

Data on career approximate value (AV) of all National Football League (NFL) players drafted between 2002 and 2014 show how career AV varies by round, position, and team. Career AVs of the drafted players are then used to assess winners and losers in each of the thirteen annual NFL drafts. Finally, regression analysis is employed to assign a trade value to each of the 224 players picked in a 32-team, seven-round draft. These values are used to evaluate the draft day trades in 2010 and which teams stand to benefit from trades up to and during the 2015 NFL draft.

Introduction

Reverse-order drafts, in theory, allocate the best new players to the weakest teams. In the National Football League (NFL) Annual Selection, better known as the NFL draft, all thirty-two NFL teams[1] pick the players they value the highest. The first twenty picks in each round are given to teams that do not make the playoffs, ordered from worst to best based on regular season records. Teams that make the playoffs are then ordered according to which round of the playoffs they are eliminated; the earlier their elimination, the higher their pick. The loser of the Super Bowl picks next to last and the winner picks last. The same order of selection continues through all seven rounds of the draft.

The logic behind the reverse-order draft is that first-round picks are more valuable than second- or later-round picks and higher first-round picks, in turn, are more valuable than lower first-round picks. Reverse-order drafts should promote competitive balance by allocating the best draft-eligible players to the weakest teams. Yet, in practice, every draft has high-profile busts and unexpected successes. There is no guarantee that any of the teams with a first-round draft pick will find a full-time starter (or that a team with a poor draft pick will fail to find a future Hall of Famer).[2]

Success in equalizing talent depends on the ability of teams to identify and develop talented players. But, how well do NFL teams identify new talent? Which NFL teams find talented players despite having poor draft picks? Which NFL teams fail to find talented players despite having excellent draft picks? And, how does the value of a draft pick vary across all seven rounds of the draft?

Evaluating talent in the NFL might seem especially difficult (if not impossible) since a nose tackle, say, has little in common with a running back or quarterback. Yet, Pro-Football-Reference.com (launched in 2003) reports a comprehensive metric modestly called "Approximate Value" (hereafter, AV) to assess player worth at any position in any year (for details, see www.pro-football-reference.com/about/glossary.htm).[3] In this paper, we compute the career AV and the AV accumulated for the team that drafted the player (hereafter, draft AV) of all 3323 players belonging to the thirteen draft classes spanning the years 2002 through 2014. The data on career AV and draft AV for each drafted player are from www.pro-football-reference.com/draft/.

The Data

Table 1 shows the average career AV and draft AV by round. Not surprisingly, the averages decline from round to round. First-round picks are regarded as the jewels of a draft class and they are expected to contribute the most. As seen in Table 1, the average career AV (draft AV) of round 1 picks is almost 55 (65) percent higher than the corresponding average of second-round picks. The average career AV (draft AV) of first-round picks is close to 5.6 (6.3) times higher than the corresponding average of seventh-round picks.

Prior to each NFL draft there is seemingly endless speculation as to which position will be most valuable and which prospects will most likely become impact players. Most fans would agree that quarterback is the most valuable position on the field. After all, quarterbacks handle just about every snap and are responsible for most on-field adjustments and decisions. While some fans believe that offensive tackle is the second most important position on the field, others believe that players who handle the ball such as running backs or wide receivers can have an even greater impact. Table 2 shows how career AV and draft AV vary by position. For the players drafted between 2002 and 2014, tackles have the highest average career AV and draft AV, followed by centers. Tackles and centers enjoy much longer careers than running backs and wide receivers, probably because offensive linemen are not subject to the same type of physical abuse that running backs and wide receivers must endure. Quarterbacks, in turn, have longer careers and hence more time to accumulate approximate value points than do running backs and wide receivers.[4] Not surprisingly, longer careers tend to increase career AVs.

Table 3 shows career AV and draft AV of draft picks by team. Between 2002 and 2014, nine different teams — New England Patriots (2002, 2004, 2005), New York Giants (2008, 2012), Pittsburgh Steelers (2006, 2009), Tampa Bay Buccaneers (2003), Indianapolis Colts (2007), New Orleans Saints (2010), Green Bay Packers (2011), Baltimore Ravens (2013), and Seattle Seahawks (2014) — won the Super Bowl. The three multiple winners — Patriots, Giants, and Steelers — rank among the top twelve teams by career AV of drafted players and among the top ten teams by draft AV.

Winners and Losers in the NFL Draft

To gauge the success of each team in each year, all teams were divided into four groups. Group 1 teams have a winning percentage above .500 and draft picks (following their "above .500" season) whose average career AV (through 2014) is above the average career AV of all players in that year's draft class. For example, the New England Patriots had a .875 winning percentage in 2002. The career AV of their 2003 draft picks (through 2014) is 26.5; the average career AV of all players drafted in 2003 is 19.755. Hence, the 2003 New England Patriots belong to Group 1. Group 2 teams have a winning percentage above .500 and draft picks thereafter whose average career AV (through 2014) is below the average career AV of all players in that year's draft class. For example, the 2013 Washington Redskins had a .625 winning percentage in 2012 and their 2013 draft picks have an average career AV (through 2014) of 2.833; the average career AV of all players drafted in 2013 is 4.786. Hence, the 2013 Redskins belong to Group 2. Group 3 teams have a winning percentage of .500 or less and draft picks thereafter whose average career AV (through 2014) is below the average career AV of all players in that year's draft class. For example, the 2005 Detroit Lions had a .375 winning percentage in 2004 and their 2005 draft picks have an average career AV (through 2014) of 8.5; the average career AV of all players drafted in 2005 is 19.326. Hence, the 2005 Lions belong to Group 3. Finally, Group 4 teams have a winning percentage of .500 or less and draft picks thereafter whose average career AV (through 2014) is above the average career AV of all players in that year's draft class. For example, the 2007 Minnesota Vikings had a .375 winning percentage in 2006 and their 2007 draft picks have a career AV (through 2014) of 22.857; the average career AV of all players drafted in 2007 is 16.540. Hence, the 2007 Vikings belong to Group 4.

Most teams should end up in either Group 2 or Group 4. That is, teams with the best records choose last and hence pick the worst players available (Group 2); teams with the worst records choose first and hence pick the best players available (Group 4). Teams in either Group 1 or Group 3 buck the trend. The "winners" ("losers") in Group 1 (Group 3) have the best (worst) records and nonetheless pick players who contribute more (less) than the average career AV of members of that year's draft class.

Tables 4 and 5 summarize the results for all teams in each of the thirteen years. Using career AV as the metric of accomplishment, the "winners" (with the highest frequency of "1" designations) include the Packers (seven years) and the Patriots (six); the "losers" (with the highest frequency of "3" designations) include the Lions (eight years), the Raiders, Redskins, Rams, and Bills (with seven each). Using draft AV as the metric of accomplishment, the "winners" and "losers" are as before, with the addition of one more notable winner (the Steelers) and one more notable loser (the Browns).

The three biggest winners in the NFL draft between 2002 and 2014 (Group 1 teams) in Table 4 with the largest z-scores[5] based on that year's draft class career AV are the 2007 New York Jets ($z = 3.363$), the 2005 New England Patriots ($z = 2.887$), and the 2006 Denver Broncos ($z = 2.641$). The three biggest losers (Group 3 teams) in Table 4 with the smallest z-scores (that is, the largest negative values) are the 2006 St. Louis Rams ($z = -1.713$), the 2002 Arizona Cardinals ($z = -1.647$), and the 2014 Detroit Lions ($z = -1.504$).[6]

A New Draft Trade Value Chart

Pro-Football-Reference.com is the source for a "draft trade value chart" (see www.pro-football-reference.com/draft/draft_trade_value.htm), devised by Jimmy Johnson (during his tenure as head coach of the Dallas Cowboys, 1989-1993). This chart assigns a value to each of the first 224 picks in the NFL draft. The number one pick in the first round is worth 3000 points, while the number 32 pick in the seventh round is worth only 2 points. According to this chart, reproduced in Table 6, pick 16 in the first round (1000 points) is considered equivalent to pick 31 in the first round (600 points) plus pick 18 in the second round (400 points).[7] The average value of round 1 picks is 1158; the average value of round 7 picks is only 8.009. Round 1 picks, on average, would appear to be over 144 times more valuable than round 7 picks. But, Table 1 indicated that the average career AV of round 1 picks is only about 5.6 times higher than the average career AV of round 7 picks. In other words, Jimmy Johnson's draft trade value chart does not appear to be a faithful representation of drafted players' career AV (as pointed out by Kevin Meers, https://harvardsportsanalysis.wordpress.com/2011/11/30/how-to-value-nfl-draft-picks/).

The average career AV of each pick number (1 through 224) covering the thirteen draft classes was computed and plotted against the pick number. The curvilinear relationship was (not surprisingly) negatively sloped. The slope decreased in absolute value as the pick number increased. To capture this nonlinearity, a regression of the average career AV [*CareerAV*] for each pick number was regressed against the natural logarithm of the pick number [*ln(pick_number)*] as follows (*t*-values in parentheses):

(1) CareerAV = 53.6728 - 8.9315 ln(pick_number)
          (37.87)   (-28.54)
                                    $R^2 = .7859$

The scatterplot relating *CareerAV* and *ln(pick_number)* is shown in Figure 1. The fit is remarkably strong with a correlation between the two variables of -0.887. Using equation (1), one can predict the career AV of overall number one picks, namely, 53.6728 – 8.9315 × ln(1) or 53.6728 (since the natural log of 1 is zero) and rescale the value of the number one pick to "3000" (by simply multiplying 53.6728 by 3000/53.6728). The value of the number two pick in the first round would be

2654 {= [53.6728 – 8.9315 × ln(2)] × (3000/53.6728)}.

All values of the remaining 222 picks are reported in Table 7.

As of May 2015, the NFL reported details of all trades on its 2015 NFL draft trade tracker (see http://www.nfl.com/news/story/0ap3000000481429/article/2015-nfl-draft-trade-tracker-details-of-all-the-moves). Each of the eighteen trades (involving one team's picks for another team's picks) are reported in Table 8. Each trade is evaluated using Jimmy Johnson's chart values and those reported in Table 7. If one team's cumulative value is within ten percent of the other team's cumulative value then the trade is considered an even exchange. Using Jimmy Johnson's chart, thirteen of the eighteen trades are "even," suggesting that NFL teams closely follow Jimmy Johnson's chart values. Using the values of Table 7, only one of the eighteen trades would be considered an "even" exchange. Table 7 ascribes much more value to picks in later rounds (than does Jimmy Johnson) and, as a result, trades tend to favor teams that acquire multiple picks deep into the draft. For example, Jimmy Johnson would regard the trade between the Redskins and the Seahawks (seventh line from the bottom of Table 8) as fair to both teams (that is, point values for each team — 233.6 and 245 — are within ten percent of each other). The values in Table 7 suggest that the Redskins will benefit immensely. Of course, only time will tell which team gets the better of the 2015 trades.

Tables 6 and 7 can also be used to evaluate past trades involving draft picks. The 2010 NFL draft trade tracker (http://sports.espn.go.com/nfl/draft10/news/story?id=5118919) reports twenty-two trades involving exclusively picks from that year's draft class. These twenty-two trades are shown in Table 9, along with their point values. If the cumulative point value of the picks received by one team is within ten percent of the corresponding point value of the other team (using either Jimmy Johnson's values in Table 6 or the new chart in Table 7), then there is no clear "winner." If, however, point totals are more than ten percent higher for one team, then that team is identified as the "winner." Career AVs (through the 2014 season) for members of the 2010 draft class involved in these trades are reported in square brackets in Table 9. Jimmy Johnson's chart would have correctly called five clear winners (four from Day 3 and one from Day 2); Table 7 would have correctly called eleven winners (three from Day 3; four each from Days 2 and 1). There are notable misses using Jimmy Johnson's chart. On Day 2, the Browns received pick number 59 (Jimmy Johnson points: 310; Table 7 points: 964); the Eagles received pick numbers 71, 134, and 146 (Jimmy Johnson points: 235 + 39 + 33 = 307; Table 7 points: 872 + 555 + 512 = 1939). The career AV of pick number 59 is 9; the cumulative career AV of pick numbers 71, 134, and 146 is 21 + 1 + 14 or 36. The Eagles appear to have "won" that trade, as would have been predicted by Table 7. For another example, on Day 2, the Buccaneers received pick number 39 (Jimmy Johnson points: 510; Table 7 points: 1171); the Raiders received in exchange pick numbers 42 and 153 (Jimmy Johnson points: 480 + 30.2 = 510.2; Table 7 points: 1134 + 489 = 1623). The Raiders then (later on Day 2) traded away their just acquired pick number 42 to the Patriots (Jimmy Johnson points: 480; Table 7 points: 1134); the Raiders received pick numbers 44 and 190 (Jimmy Johnson points: 460 + 15.4 = 475.4; Table 7 points: 1111 + 381 = 1492).[8] This trade between the Raiders and the Patriots would be, using Jimmy Johnson's point values, a fair exchange (480 vs. 475.4). Using Table 7, the Raiders should never have made this trade with the Patriots because the point value (1492) was *less than* what they obtained (1623) when they first acquired pick number 42. The Raiders' fate might have been different if they had never traded away pick number 42 in the 2010 draft, tight end Rob Gronkowski.

Concluding Remarks

The reverse-order NFL draft is subject to change as teams execute trades up to and during each year's draft. Until now, the value of acquired picks has been based on a chart developed back in the 1990s by former Cowboys head coach Jimmy Johnson. This chart does not accurately reflect players' approximate value, a metric used to assess player worth at any position in any year. A chart that does ascribe value based on the 13-year (2002 through 2014) average career approximate value for each of the 224 picks in the NFL draft shows that players drafted in later rounds are grossly undervalued. Thus packaging multiple lower round picks in exchange for a single high round pick may vault also-rans into elite company.

Table 1
Career AV and Draft AV, All Drafted Players 2002 – 2014, Averages by Round

Round	Career AV[a]	Draft AV[b]
1	31.484	27.036
2	20.327	16.394
3	13.744	10.915
4	11.475	9.224
5	8.352	6.417
6	6.378	4.836
7	5.627	4.299

[a]Weighted career approximate value (AV) found by summing 100 percent of the player's AV in his best season, 95 percent of his AV in his second-best season, and so forth.
[b]The AV accumulated for the team that drafted this player.

Table 2
Career AV and Draft AV, All Drafted players 2002 – 2014, Averages by Position

Position	Career AV	Rank	Draft AV	Rank
Center	18.701	2	16.410	2
Defensive Back	12.439	9	10.232	9
Defensive End	16.173	4	13.372	3
Defensive Tackle	13.430	7	11.293	8
Fullback	3.267	14	2.321	14
Guard	17.442	3	13.366	4
Kicker	9.115	12	7.800	12
Linebacker	14.033	6	11.903	7
Punter	9.520	11	8.000	11
Quarterback	15.611	5	12.215	5
Running Back	13.157	8	12.000	6
Tackle	19.044	1	16.923	1
Tight End	8.495	13	6.994	13
Wide Receiver	11.901	10	10.114	10

Table 3
Career AV and Draft AV, All Drafted Players 2002 – 2014, Averages by Team

Team	Career AV	Rank	Draft AV	Rank
Arizona Cardinals (ARI)	14.854	7	12.157	9
Atlanta Falcons (ATL)	14.351	11	11.756	13
Baltimore Ravens (BAL)	14.308	13	11.636	14
Buffalo Bills (BUF)	12.019	27	10.082	26
Carolina Panthers (CAR)	15.302	5	12.682	7
Chicago Bears (CHI)	14.484	10	11.393	17
Cincinnati Bengals (CIN)	12.951	23	11.062	20
Cleveland Browns (CLE)	12.533	26	9.591	28
Dallas Cowboys (DAL)	15.552	4	13.138	6
Denver Broncos (DEN)	13.947	16	10.512	22
Detroit Lions (DET)	11.287	29	9.352	30
Green Bay Packers (GNB)	14.573	9	13.490	5
Houston Texans (HOU)	13.327	17	11.400	16
Indianapolis Colts (IND)	14.042	15	11.223	19
Jacksonville Jaguars (JAX)	14.053	14	11.933	12
Kansas City Chiefs (KAN)	13.097	21	10.319	23
Miami Dolphins (MIA)	12.559	25	10.141	25
Minnesota Vikings (MIN)	13.191	19	12.114	11
New England Patriots (NEW)	14.317	12	13.517	4
New Orleans Saints (NOR)	16.355	2	15.176	2
New York Giants (NYG)	15.230	6	12.145	10
New York Jets (NYJ)	14.726	8	12.231	8
Oakland Raiders (OAK)	11.546	28	8.778	31
Philadelphia Eagles (PHI)	13.138	20	11.432	15
Pittsburgh Steelers (PIT)	15.804	3	15.272	1
San Diego Chargers (SDG)	17.930	1	14.618	3
San Francisco 49ers (SFO)	13.284	18	11.039	21
Seattle Seahawks (SEA)	13.029	22	11.392	18
St. Louis Rams (STL)	10.717	30	8.434	32
Tampa Bay Buccaneers (TAM)	10.402	31	9.700	27
Tennessee Titans (TEN)	12.750	24	10.171	24
Washington Redskins (WAS)	10.163	32	9.432	29

Table 4
Group Designations Using Career AV, by Team

Team	2002	2003	2004	2005	2006	2007	2008	2009	2010	2011	2012	2013	2014
ARI	3	4	4	3	3	4	4	2	1	4	3	3	2
ATL	3	2	4	1	3	4	4	2	2	1	2	2	3
BAL	1	4	2	2	4	1	4	1	2	2	1	2	4
BUF	3	4	3	2	3	4	3	4	3	4	3	3	4
CAR	4	3	1	4	1	4	4	2	3	4	4	4	1
CHI	2	4	4	4	2	2	3	2	3	1	3	1	3
CIN	3	3	3	3	1	3	3	4	2	4	1	2	1
CLE	3	1	3	3	3	4	2	3	4	4	4	3	4
DAL	4	4	2	4	2	2	1	2	1	4	3	4	4
DEN	3	2	2	2	1	2	4	3	4	4	4	2	2
DET	3	3	3	3	3	4	3	4	4	3	2	4	3
GNB	1	2	1	1	4	3	1	4	1	2	2	1	1
HOU	·	3	3	3	4	3	4	4	2	4	1	2	3
IND	4	1	2	2	1	2	2	1	2	2	4	2	1
JAX	4	4	4	2	1	4	2	4	3	3	3	4	4
KAN	3	3	1	3	2	2	4	3	4	2	3	4	1
MIA	2	2	2	4	2	4	4	2	4	4	4	3	4
MIN	4	4	2	3	1	4	4	1	2	3	4	2	4
NOR	3	2	3	3	4	2	4	3	1	2	2	4	2
NWE	1	1	2	1	2	2	2	1	1	2	1	2	2
NYG	3	1	4	4	2	3	2	1	3	2	2	2	4
NYJ	1	2	4	2	3	1	4	1	2	1	3	4	3
OAK	1	2	3	3	4	3	4	4	3	3	3	3	4
PHI	1	2	2	1	4	2	4	1	2	2	4	4	2
PIT	1	1	4	2	2	4	2	1	1	2	2	4	3
SDG	4	3	4	1	1	1	2	3	1	2	3	4	2
SEA	2	3	2	1	2	2	2	3	4	4	4	2	2
SFO	2	2	3	4	3	4	3	3	4	3	2	2	2
STL	2	3	1	3	3	3	3	4	3	3	4	4	4
TAM	2	2	3	3	2	3	1	1	3	2	4	4	3
TEN	3	2	2	4	3	3	1	2	3	4	2	4	4
WAS	3	3	3	4	1	3	2	3	4	3	4	2	3

Table 5
Group Designations Using Draft AV, by Team

Team	2002	2003	2004	2005	2006	2007	2008	2009	2010	2011	2012	2013	2014
ARI	3	4	4	3	3	4	4	2	1	4	3	3	2
ATL	3	2	4	1	3	3	4	2	2	1	2	2	3
BAL	1	4	2	2	4	1	4	1	2	2	1	2	4
BUF	3	4	3	2	3	3	3	3	3	4	3	3	4
CAR	4	3	1	3	2	4	3	2	3	4	4	4	1
CHI	2	4	4	3	2	2	4	2	3	2	4	1	3
CIN	3	3	3	3	1	3	3	4	1	4	2	2	1
CLE	3	2	3	3	3	4	2	3	4	4	3	3	4
DAL	4	4	2	4	2	2	1	2	1	4	3	4	4
DEN	3	2	2	2	1	2	3	3	4	4	4	2	2
DET	3	3	3	3	3	3	3	4	4	3	2	4	3
GNB	1	2	1	1	4	3	1	4	1	2	2	1	1
HOU	·	4	3	3	4	3	4	4	2	4	1	2	3
IND	4	1	2	2	1	2	2	1	2	2	4	2	1
JAX	4	4	4	2	1	3	2	3	3	3	3	4	4
KAN	3	3	2	4	2	2	4	3	4	2	3	4	1
MIA	2	2	1	4	2	3	4	2	4	4	4	3	4
MIN	4	4	2	3	1	4	4	1	2	3	4	2	4
NOR	4	2	4	3	4	1	4	4	1	2	2	4	2
NWE	1	1	2	1	2	2	2	1	1	2	1	2	1
NYG	3	1	4	4	2	3	1	1	3	2	2	2	4
NYJ	1	2	3	2	4	1	3	1	2	1	4	4	3
OAK	2	2	3	3	3	3	4	3	3	3	3	3	4
PHI	1	2	2	1	4	2	3	1	2	2	4	4	2
PIT	1	1	4	1	1	4	2	1	1	2	2	4	3
SDG	4	3	4	1	1	2	2	4	1	2	3	4	2
SEA	2	3	2	1	2	2	2	3	4	4	4	2	2
SFO	2	2	3	4	3	4	3	3	4	4	2	2	2
STL	2	3	1	3	3	3	3	4	3	4	4	4	4
TAM	2	2	3	3	2	3	2	1	3	2	4	4	3
TEN	3	2	2	4	3	3	1	2	3	4	2	4	4
WAS	3	3	4	3	1	4	2	3	4	3	4	2	3

Table 6
NFL Draft Trade Value Chart

Pick number	Round						
	1	2	3	4	5	6	7
1	3000	580	265	112	43	27	14.2
2	2600	560	260	108	42	26.6	13.8
3	2200	550	255	104	41	26.2	13.4
4	1800	540	250	100	40	25.8	13
5	1700	530	245	96	39.5	25.4	12.6
6	1600	520	240	92	39	25	12.2
7	1500	510	235	88	38.5	24.6	11.8
8	1400	500	230	86	38	24.2	11.4
9	1350	490	225	84	37.5	23.8	11
10	1300	480	220	82	37	23.4	10.6
11	1250	470	215	80	36.5	23	10.2
12	1200	460	210	78	36	22.6	9.8
13	1150	450	205	76	35.5	22.2	9.4
14	1100	440	200	74	35	21.8	9
15	1050	430	195	72	34.5	21.4	8.6
16	1000	420	190	70	34	21	8.2
17	950	410	185	68	33.5	20.6	7.8
18	900	400	180	66	33	20.2	7.4
19	875	390	175	64	32.6	19.8	7
20	850	380	170	62	32.2	19.4	6.6
21	800	370	165	60	31.8	19	6.2
22	780	360	160	58	31.4	18.6	5.8
23	760	350	155	56	31	18.2	5.4
24	740	340	150	54	30.6	17.8	5
25	720	330	145	52	30.2	17.4	4.6
26	700	320	140	50	29.8	17	4.2
27	680	310	136	49	29.4	16.6	3.8
28	660	300	132	48	29	16.2	3.4
29	640	292	128	47	28.6	15.8	3
30	620	284	124	46	28.2	15.4	2.6
31	600	276	120	45	27.8	15	2.3
32	590	270	116	44	27.4	14.6	2

Source: http://www.pro-football-reference.com/draft/draft_trade_value.htm

Table 7
Revised NFL Draft Trade Value Chart

Pick number	Round						
	1	2	3	4	5	6	7
1	3000	1254	916	716	574	463	373
2	2654	1240	908	711	570	460	370
3	2452	1225	901	706	566	457	368
4	2308	1211	894	701	562	454	365
5	2197	1197	886	696	559	451	363
6	2106	1184	879	691	555	448	360
7	2029	1171	872	686	551	445	357
8	1962	1158	865	681	548	442	355
9	1903	1146	858	677	544	439	352
10	1851	1134	851	672	540	436	350
11	1803	1122	845	667	537	433	348
12	1759	1111	838	663	533	430	345
13	1720	1100	831	658	529	427	343
14	1683	1089	825	653	526	425	340
15	1648	1078	819	649	522	422	338
16	1616	1067	812	644	519	419	335
17	1586	1057	806	640	516	416	333
18	1557	1047	800	636	512	413	331
19	1530	1037	794	631	509	410	328
20	1504	1027	788	627	505	408	326
21	1480	1018	782	623	502	405	324
22	1457	1009	776	618	499	402	321
23	1435	999	771	614	495	399	319
24	1413	990	765	610	492	397	317
25	1393	982	759	606	489	394	314
26	1373	973	754	602	485	391	312
27	1355	964	748	598	482	389	310
28	1337	956	743	594	479	386	307
29	1319	948	737	590	476	383	305
30	1302	940	732	586	473	381	303
31	1286	932	727	582	470	378	301
32	1270	924	721	578	466	375	298

Table 8
Estimated Value of Selected 2015 NFL Draft Trades[1]

	Point value			Point value	
	Table 6	Table 7		Table 6	Table 7
Colts receive: No. 151	**31**[a]	495	49ers receive: Nos. 165, 244[b]	27.4	**707**
Packers receive: No. 147	**32.6**	509	Patriots receive: Nos. 166, 247	27	**698**
Falcons receive: No. 137	37.5	544	Vikings receive: Nos. 146, 185	**50.4**	**906**
Buccaneers receive: No. 124	48	594	Raiders receive: Nos. 128, 218	48.2	**890**
Cardinals receive: No. 116	62	627	Browns receive: Nos. 123, 198, 241	63.2	**1220**
Jets receive: No. 103	88	686	Jaguars receive: Nos. 104, 229	88	**968**
Panthers receive: No. 102	**92**	691	Raiders receive: Nos. 124, 161, 242	77	**1317**
Patriots receive: Nos. 111, 147, 202	115.2	**1508**	Browns receive: Nos. 96, 219	119.8	1031
Lions receive: No. 80	190	812	Vikings receive: Nos. 88, 143	184.5	**1287**
Chiefs receive: No. 76	210	838	Vikings receive: Nos. 90, 193	204.2	**1185**
Jets receive: Nos. 82, 152, 229	212.6	**1579**	Texans receive: No. 70	**240**	879
Redskins receive: Nos. 95, 112, 167, 181	233.6	**2221**	Seahawks receive: No. 69	245	886
Colts receive: Nos. 65, 109	341	1574	Buccaneers receive: Nos. 61, 128	336	1526
Cardinals receive: Nos. 58, 158	348.2	**1446**	Ravens receive: No. 55	350	999
Eagles receive: Nos. 47, 191	445	1456	Dolphins receive: Nos. 52, 145, 156	442.5	**2022**
Browns receive: Nos. 51, 116, 195	465.4	**2032**	Texans receive: Nos. 43, 229	472	1409
Panthers receive: No. 41	490	1146	Rams receive: Nos. 57, 89, 201	486	**2093**
Giants receive: No. 33	580	1254	Titans receive: Nos. 40, 108, 245	580	**2075**

[1]All of the trades reported here are from http://www.nfl.com/news/story/0ap3000000481429/article/2015-nfl-draft-trade-tracker-details-of-all-the-moves .

[a]Numbers in boldface are at least 10 percent higher than the point value of the other team involved in the same trade (using the same draft trade value chart, either Table 6 or Table 7).

[b]Selections that involve pick numbers above 224 are assigned (i) a point value of 2, using Table 6 and (ii) a point value of $\{[53.6728 - 8.9315 \ln(\text{pick\_number})] \times (3000/53.6728)\}$, using Table 7.

Table 9
Estimated Value of Selected 2010 NFL Draft Trades[1]

	Point value			Point value	
	Table 6	Table 7		Table 6	Table 7

Day 3

Patriots receive: No. 208	**8.2**[a]	335 [2][b]	Redskins receive: Nos. 229, 231[b]	4	**570** [0]
Dolphins receive: No. 163	26.2	457 [25]	Redskins receive: Nos. 174, 219	25.6	**735** [1]
Falcons receive: No. 135	38.5	551 [6]	Rams receive: Nos. 144, 189	**49.8**	902 [8]
Dolphins receive: No. 119	56	614 [0]	Cowboys receive: Nos. 126, 179	**65.8**	996 [6]
Saints receive: No. 123	49	598 [6]	Cardinals receive: Nos. 130, 201	53	**918** [13]
Jets receive: No. 112	**86**	681 [7]	Panthers receive: Nos. 124, 198	60.2	**954** [5]

Day 2

Chiefs receive: No. 93	128	737 [10]	Texans receive: Nos. 102, 144	126	**1210**[14]
Packers receive: No. 71	**235**	872 [21]	Eagles receive: Nos. 86, 122	210	**1378**[12]
Browns receive: No. 59	310	964 [9]	Eagles receive: Nos. 71, 134, 146	307	**1939**[36]
Texans receive: No. 58	320	973 [14]	Patriots receive: Nos. 62, 150	315.4	**1439**[28]
Cowboys receive: No. 55	350	999 [17]	Eagles receive: Nos. 59, 125	357	**1554**[10]
Vikings receive: No. 51	390	1037[16]	Texans receive: Nos. 61, 93	420	**1658**[15]
Cardinals receive: No. 47	430	1078[36]	Patriots receive: Nos. 58, 89	465	**1732**[15]
Patriots receive: No. 42	480	1134[43]	Raiders receive: Nos. 44, 190	475.4	**1492**[31]
Buccaneers receive: No. 39	510	1171 [7]	Raiders receive: Nos. 42, 153	510.2	**1623**[50]

Table 9
Estimated Value of Selected 2010 NFL Draft Trades[1]
(Continued)

	Point value			Point value	
	Table 6	Table 7		Table 6	Table 7

Day 1

49ers receive: No. 11	1250	1803[35]	Broncos receive: Nos. 13, 113	1218	**2360**[31]
Chargers receive: Nos. 12, 110, 173	1296.2	2839[49]	Dolphins receive: Nos. 28, 40, 126	1206	3081[47]
Eagles receive: No. 13	1150	1720 [9]	Broncos receive: Nos. 24, 70, 87	1135	**3063**[93]
Broncos receive: No. 22	780	1457[46]	Patriots receive: Nos. 24, 113	808	**2053**[70]
Cowboys receive: Nos. 24, 119	796	2027[48]	Patriots receive: Nos. 27, 90	820	2109[34]
Broncos receive: No. 25	720	1393[12]	Ravens receive: Nos. 43, 70, 114	776	**2637**[22]
Lions receive: Nos. 30, 128	664	1880[17]	Vikings receive: Nos. 34, 100, 214	665.8	**2262**[23]

[1]All of the trades reported here are from http://sports.espn.go.com/nfl/draft10/news/story?id=5118919 .
[a]Numbers in boldface are at least 10 percent higher than the point value of the other team involved in the same trade (using the same draft trade value chart, either Table 6 or Table 7).
[b]Numbers in square brackets are career AVs of the acquired pick(s) through the 2014 season.

Figure 1
Scatterplot of Career AV versus ln(pick number)

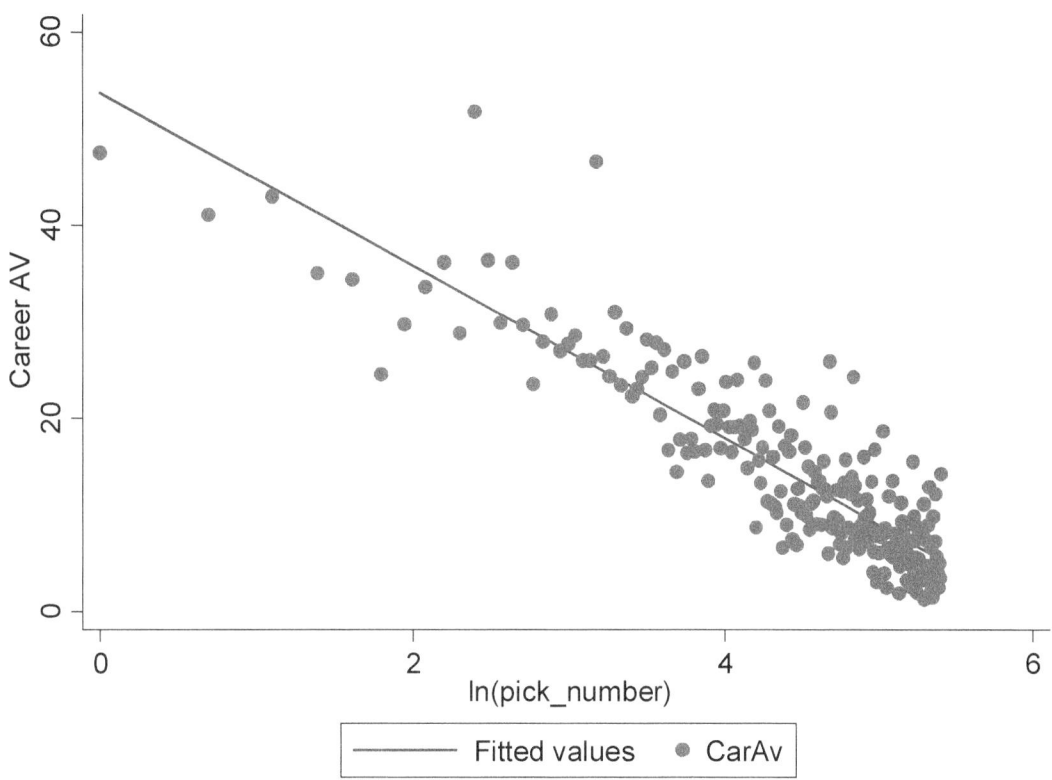

References

Approximate value. (n.d.). Retrieved from

    www.pro-football-reference.com/about/glossary.htm

Draft classes. (n.d.). Retrieved from

    www.pro-football-reference.com/draft/

Draft trade value chart. (n.d.). Retrieved from

    www.pro-football-reference.com/draft/draft_trade_value.htm

Green, C. E. and Sommers, P. M. (2007). "Is a defensive tackle worth more than a quarterback?" **J. of Recreational Mathematics,** 36(3), 233-239.

Meers, K. (2011). How to value NFL draft picks. Retrieved from

    https://harvardsportsanalysis.wordpress.com/2011/11/30/how-to-value-nfl-draft-picks/

2010 NFL draft trade tracker. (2010). Retrieved from

    http://sports.espn.go.com/nfl/draft10/news/story?id=5118919

2015 NFL draft trade tracker. (2015). Retrieved from

http://www.nfl.com/news/story/0ap3000000481429/article/2015-nfl-draft-trade-tracker-details-of-all-the-moves

Footnotes

1. In 2002, the NFL expanded to thirty-two teams, with the addition of the Houston Texans.

2. In 2000, for example, the New England Patriots chose future Hall of Famer quarterback Tom Brady in the sixth round of the draft.

3. Approximate value was created by mathematician Doug Drinen. Pro-Football-Reference.com also reports a "Weighted Career AV" (hereafter, career AV) for each player, which is the sum of 100 percent of the player's AV in his best season, plus 95 percent of his second-best season, plus 90 percent of his third-best season, and so forth. The career AV can be used to compare career productivity of players at different positions. Less than half (1574) of the 3323 players drafted between 2002 and 2014 had retired before 2014. So, the career AVs used here represent "snapshots" at a point in time, namely, career totals as of the end of the 2014 season.

4. See Green and Sommers (2007) for a more detailed analysis of the longevity and relative worth of each position.

5. A standardized score is a z-score, that is, the number of standard deviations above the mean (if the z-score is positive) or below the mean (if the z-score is negative).

6. Using draft AV (Table 5), the three biggest winners are the 2002 Philadelphia Eagles ($z = 2.958$), the 2007 New York Jets ($z = 2.845$), and the 2005 New England Patriots ($z = 2.777$); the three biggest losers are the 2014 Detroit Lions ($z = -1.605$); the 2014 Houston Texans ($z = -1.553$), and the 2006 St. Louis Rams ($z = -1.549$).

7. The website incorrectly reports "pick 28 in the second round," which would be the 60th overall pick with a value of only 300 (not 400) points.

8. Neither Jimmy Johnson nor Table 7 would have correctly predicted a "winner" for the Day 2 trade between the Raiders and the Patriots. Jimmy Johnson values the exchange equally (480 points *vs.* 475.4 points); Table 7 would have given the edge to the Raiders (1134 points *vs.* 1492 points). But, as mentioned in the text, the Raiders should not have given up pick number 42 by trading *down* from 1623 to 1492 points.

Wordplay Sayings Translated

Penny for your thoughts?

Actions speak louder than words.

To add insult to injury…

Back to the drawing board.

It's the best thing since sliced bread.

You've bitten off more than you can chew.

It costs an arm and a leg.

Drastic times call for drastic measures.

Every cloud has a silver lining.

It takes two to tango.

A picture paints a thousand words.

Endure it with a granule of sodium chloride.

Black holes result from God dividing the universe by zero. – Author unknown

Infinity is a floorless room without walls or ceiling. - Author Unknown

God is real, unless declared integer. ~Author Unknown

Solution to Mathematicians in the "School of Athens"

```
            E
            P
            I
            C
            U           A
            R           R
        E   U   C   L   I   D           P
            S           S               T
                P   Y   T   H   A   G   O   R   A   S
                        O               L
                        T       D       E
        H               L       I       M
        E           Z   E   N   O       Y
        R               O       G
        A   V   E   R   R   O   E   S
        C           O           N           X
        L       P   A   R   M   E   N   I   D   E   S
        I           S           S           N
        T           T                       O
        U           E                       P
        S   O   C   R   A   T   E   S       H
                                            O
                                            N
```

81

BOOK REVIEWS

Edited by:Charles Ashbacher

Charles Ashbacher Technologies

5530 Kacena Ave

Marion, IA 52302

E-mail: cashbacher@yahoo.com

Brain Puzzle Game, 3D Magical Intellect Ball Kids Toys Gift Intelligence Toys, sold online by SoFu. $9.99

At approximately six inches in diameter and with a ridge that extends out approximately one-quarter of an inch on the equator, this spherical three-dimensional puzzle is easy to hold on to and manipulate. This is critical, for it requires a set of steady hands to move the ball through all phases of the maze.

It is constructed from a set of tracks inside the sphere where the goal is to cause the ball to transverse from the start to the end point. Getting it into one of the three start positions is fairly easy, there is a red structure that is attached to the side. The start position opens to the yellow-colored plane of tracks, the green-colored plane is perpendicular to it and the blue-colored plane has two sections that are perpendicular to the other two while one section is tilted at approximately a sixty degree angle.

The ridges on the track are generally very low, so it is hard to keep the ball from dropping off of it. One very interesting characteristic is that the ball can travel on both sides of the tracks. This increases the distance that the ball must cover before you can declare success.

Solving this puzzle requires the player to engage in some three-dimensional thinking and planning. You must always be looking ahead and be ready to tilt it when you reach the end of a track in one of the dimensions. It can keep you occupied for hours.

Charles Ashbacher

Proofs Without Words III: Further Exercises in Visual Thinking, by Roger B. Nelsen, The Mathematical Association of America, Washington, D. C., 2016. 187 pp., $50.00 (paper). ISBN 978-0-88385-790-8.

While some may quibble about whether these images are rigorous mathematical proofs, none can dispute their power and elegance. Nelsen has once again captured the essence of mathematics in a series of easy to understand, yet thorough diagrams illustrating fundamental mathematical concepts.

 The proofs are organized into five categories. They are:

*) Geometry & Algebra

*) Trigonometry, Calculus & Analytic Geometry

*) Inequalities

*) Integers & Integer Sums

*) Infinite Series & Other Topics

 Instructors of courses covering this content will most likely find a diagram that they can use to visually demonstrate a concept. This could be of enormous help to some students that are experiencing difficulties. Personally, I have had several math students that were struggling come to me and say that they are "visual learners." Which is of course true, humans naturally understand images better than thoughts expressed in symbols.

Charles Ashbacher

Elements of Mathematics: From Euclid to Gödel, by John Stillwell, Princeton University Press, Princeton, New Jersey, 2016. 440 pp., $39.9 (hardbound). ISBN 978-0-691-17168-5.

 For those experienced in mathematics, the most interesting feature of this book is the attempt to keep things at the elementary level. As is demonstrated in many ways and emphasized by Stillwell, the phrase "elementary mathematics" is one subject to a wide variety of interpretations. Both over time as well as from person to person. What was considered advanced when introduced has become routine over the centuries.

 Yet, Stillwell does point to one concept that can be used to separate the elementary from the advanced, and that is the idea of infinity. It is an idea that will always remain abstract and requires thought processes that can accept what appears to be paradoxical. For example, the idea that the infinity of the even integers can be contained in the integers and yet they can be considered to be the same "size." Or that the infinity of the real numbers is "larger" than that of the natural numbers.

 Since the coverage begins with Euclid and flows through the centuries until the twentieth, this book is first and foremost a popular history of mathematics. Yet, there is an important underlying theme, that there is a concept that can be used to determine the difference between elementary and advanced mathematics. While it is of course not completely effective as a separator, it is a good first approximation.

This is a book that would serve well as a textbook for a liberal arts course in mathematics. There is no sparing of the formulas, some sections would have to be skipped or subject to deep explanations, yet the coverage of the fundamentals of mathematics is sufficient to justify its use.

Charles Ashbacher

The Heart of Calculus: Explorations and Applications, by Philip M. Anselone and John W. Lee, The Mathematical Association of America, Washington, D. C., 20-15. 245 pp., $60.00 (hardbound). ISBN 9780883857878.

The content of the book is such that no one can dispute the importance of the topics, just their relative importance and whether they truly should be considered within the "heart" of calculus. The first six chapters are definitely within the core of the subject. They are:

*) Critical Points and Graphing

*) Inverse Functions

*) Exponential and Logarithmic Functions

*) Linear Approximation and Newton's Method

*) Taylor Polynomial Approximations

*) Global Extreme Values

It is the content of the chapters after this where honest disagreement can and will exist. The chapter titles are:

*) Angular Velocity and Curvature

*) π and e are Irrational

*) Hanging Cables

*) The Buffon Needle Problem

*) Optimal Location

*) Energy

*) Springs and Pendulums

*) Kepler's Laws of Planetary Motion

*) Newton's Law of Universal Gravitation

*) From Newton to Kepler and Beyond

As you can see, the last ten chapters deal with major advancement achieved via the use of calculus. All of them are worthy of being included in a course in the history of mathematics where the developments are worked through as well as described. The chapters conclude with the two sections "Problems and Remarks" and "Further Reading and Projects," providing many tracks for future examination of the topic.

Calculus has such a wide spectrum of applicability that only the tiniest of fractions of the subject can be squeezed into a book of this size. If you disagree with the topic selection, just consider it a demonstrate of how much heart calculus has.

Charles Ashbacher

Cyclone Boys 50mm Cyclone Boys Speed Cube 2x2x2 Stickerless Magic Cube Puzzles

This reduced version of the classic Rubik's cube puzzle is ideal for the child that needs to be mentally and spatially challenged. It is a 2 x 2 cube rather than the classic 3 x 3 version, so it is much easier to solve. It is also large, making it easy to grasp and manipulate. The colors are also bright and attractive, enhancing the experience as the player works through the permutations in search for a solution. With so many possible configurations, if a child solves it, a parent can quickly create a new permutation and hand it back to the child so they can solve it. A great toy for extended car rides.

Charles Ashbacher

The G. H. Hardy Reader, edited by Donald J. Albers, Gerald L. Alexanderson and William Dunham, a joint publication of The Mathematical Association of America, Washington, D. C. and Cambridge University Press, Cambridge, United Kingdom. 410 pp., $49.99 (paper). ISBN 9781107594647.

Godfrey Harold (G. H.) Hardy was half of what some considered to be the three greatest mathematicians of the first half of the twentieth century, Hardy, Littlewood and Hardy/Littlewood. Hardy was a first rate mathematician and also considered to be a superb author, his most famous work is "A Mathematician's Apology," written late in life after his math skills had faded. Published in 1940 as World War II was reaching full fury, it remains one of the best explanations of the world of mathematicians as well as a human description of a professional past his prime looking back on his career. That aspect of the book can be applied to everyone from welders to physicians.

Yet, despite all of his personal achievements, Hardy's greatest contribution to mathematics was his reaction to an unsolicited letter from an unknown Indian man named Srinivasa Ramanujan. When asked by Paul Erdős what his greatest accomplishment was, Hardy did not hesitate in responding that it was his "discovery" of Ramanujan.

Given the breadth and depth of Hardy's work, the emphasis here is on the breadth with a few points delving deeper into the depth. While there are some short pieces containing advanced mathematics, most of the content can be understood by the general undergraduate. What makes this book worthy of being used in a multi-discipline liberal arts course is that it is about the man and not so much about the math. It shows a mathematician that was very much a man of the world, even though he rejected much of it, specifically religion.

This is a book that demonstrates the essence of Hardy that made him someone that was willing to take a serious look at a letter from the British colony of India. Rather than adopting the attitude that colonial subjects could have nothing meaningful to say, Hardy read it and acted on it. That one act demonstrates that at times the way to be great is to be humble.

Charles Ashbacher

Solutions Column, Journal of Recreational Mathematics Problems 2880 – 2889 From 38(2)

Edited by

Lamarr Widmer

Messiah College,
Suite 3041, One College Avenue,
Mechanicsburg, PA 17055

widmer@messiah.edu

When **Journal of Recreational Mathematics** ceased publication in December of 2014, the last issue was 38(2). A problem column appeared in that issue and readers submitted solutions to the problem editor Lamarr Widmer in the hope that they would be acknowledged. Lamarr has edited those solutions into this column and this should be the last item from **JRM** that will be published in the **TRM** series.

2880. Prime Search by Bill Dean, Wood-Ridge, New Jersey (*JRM 38*:2, p. 135)

The ordered set of odd numbers beginning with the integer 3 is a list that contains all the prime numbers greater than 2. It has only half as many integers as does the ordered set of all integers beginning at 3. One process for generating the elements of the list of odd numbers is to begin with the number 3 and repeatedly add 2.

Define a similar process that generates a list of integers that contains all the prime numbers beginning with prime P and has fewer than one-fourth as many integers as the ordered set of all integers beginning at P. As in the example given above, your list may contain integers which are not prime.

Solution by Richard Hess

We consider the list of all numbers n, greater than P such that n is congruent modulo 210 to one of the following:

$\pm 1, \pm 11, \pm 13, \pm 17, \pm 19, \pm 23, \pm 29, \pm 31, \pm 37, \pm 41, \pm 43, \pm 47, \pm 53, \pm 59, \pm 21, \pm 67, \pm 71, \pm 73, \pm 79, \pm 83, \pm 89, \pm 97, \pm 101, \pm 103$.

This list includes 48 numbers out of each set of $210 = 2 \times 3 \times 5 \times 7$ consecutive integers and includes all primes greater than P. It includes $\frac{48}{210} \approx 0.22858$ of all integers greater than P. Letting $N_k = 2 \times 3 \times 5 \times \ldots \times p_k$ and f_k be the fraction of integers included in the resulting list from the procedure outlined above, we find the data of table 1.

Table 1

k	N_k	f_k
4	210	$\frac{48}{210} \approx 0.22858$
5	2310	$\frac{16}{77} \approx 0.20779$
6	30030	$\frac{192}{1001} \approx 0.191808$
7	510510	$\frac{3072}{17017} \approx 0.180525$
8	9699690	$\frac{55296}{323323} \approx 0.171024$

In fact, $f_k = \frac{1}{2} \times \frac{2}{3} \times \frac{4}{5} \times \ldots \times \frac{p_k - 1}{p_k}$ approaches 0 as $k \to \infty$ because $\sum \frac{1}{p_k}$ diverges.

2881. Box Origami by Hubert Hagadorn, Menlo Park, CA (*JRM 38*:2, p. 135)

A box having different length, width and height is flattened after cutting a number of its edges. Assuming that the resultant planar polygon is connected, how many different shapes are possible?

Solution by the Proposer

There are 96 different shapes. They are shown in the figures below.

A cursory examination of this problem led to the conclusion that always seven edges must be cut. Since there are 12 edges, the number of possible combinations n is

$$n = \binom{12}{7} = 792.$$

Each cut combination was represented by an array consisting of 5 zeros and 7 ones. Because of symmetry considerations only one-quarter of these arrays, or 198 result in a different shape. Some combinations of cuts were eliminated because the box cannot be flattened, or the cuts result in more than one contiguous piece. To this end arrays were eliminated if there were no cut at a vertex, if all four edges of a face were cut, or if all six edges of two adjacent faces were cut.

2882. Dice Elimination by Ken Klinger, Northbrook, Il (*JRM 38*:2, p. 135)

A player rolls 10 fair six-sided dice all at one time, then selects any of the values rolled (say 5) and removes all the dice showing that value. He then rolls the remaining dice, and at each roll removes those dice showing the value chosen after the first roll (5 in our example). The goal is to remove all the dice in as few rolls as possible. What is the expected number of rolls to remove all the dice?
(Note: This is the basis of the game "Tenzi" (Copyright © 2013 Carma Games, LLC). See https://www.ilovetenzi.com/).

Solution by Richard Hess

Consider the opening roll with $d = 6^{10}$ possible outcomes, taking order into account. They can be categorized by frequencies, from greatest to least, of the 6 numbers which the dice may show. For example (6, 2, 1, 1, 0, 0) that the most common number appeared on 6 dice, the next most common two dice and two others occurred on a single die. There are 35 such patterns from (10,0,0,0,0,0) to (2,2,2,2,1,1). For each of these, we must determine the number of ways it can happen, e.g. (6,2,1,1,0,0) allows 180 choices for the dice values corresponding to frequencies 6, 2, 1 and 1. And then they can be ordered in $\frac{10!}{6! \times 2!} = 2520$ orders, giving 453,600 ways to arrive at . (6,2,1,1,0,0). Next we add the number of ways for each outcome having the same maximum frequency. For example, those with a leading 6 are (6,4,0,0,0,0), (6,3,1,0,0,0), (6,2,2,0,0,0), (6,2,1,1,0,0), (6,1,1,1,1,0). For this group of initial rolls, the player is left with four dice to continue rolling.

Hence, we denote the total probability of these five possibilities by $P_4 = \frac{787500}{d}$. Here are the probabilities, for each possible number of dice remaining after the first roll:
$P_0 = \frac{6}{d}$, $P_1 = \frac{300}{d}$, $P_2 = \frac{6750}{d}$, $P_3 = \frac{90000}{d}$, $P_4 = \frac{787500}{d}$, $P_5 = \frac{4721220}{d}$, $P_6 = \frac{18774000}{d}$, $P_7 = \frac{32004000}{d}$, $P_8 = \frac{4082400}{d}$.

Now let $p = \frac{5}{6}$ and $q = \frac{1}{6}$. Let E_m be the expected number of rolls needed to get all 10 dice to show the target number (i.e. to finish the game) from the point where m dice are not showing the showing the target number. We can get anywhere from 0 to m matches to the target number when rolling these remaining m dice.

Now
$$(p+q)^m = q^m + \binom{m}{1} q^{m-1} p + \cdots + \binom{m}{i} q^{m-i} p^i + \cdots + p^m$$

So the probability of getting i matches is $Q_i = \binom{m}{i} q^{m-i} p^i$.

Therefore $E_m = Q_0 E_m + Q_1 E_{m-1} + \cdots + Q_{m-1} E_1 + 1$. Here the final term counts the roll which brought us to the point where these probabilities apply. At the beginning of the game this is the roll which determines the target value. Now we have that $E_1 = pE_1 + 1$, which we solve to get $E_1 = 6$. And then

$E_2 = p^2 E_2 + 2pq E_1 + 1$ yields $E_2 = \frac{1+2p}{1-p^2} = \frac{96}{11}$. Continuing, we have the following.
$E_3 = \frac{2 \times 3^3 \times 587}{7 \times 11 \times 13}$, $E_4 = \frac{2^6 \times 3 \times 3793}{7 \times 11 \times 13 \times 61}$, $E_5 = \frac{2 \times 3 \times 61644183}{7 \times 11 \times 13 \times 61 \times 4651}$, $E_6 = \frac{2^5 \times 3^3 \times 32774059}{7 \times 11 \times 31 \times 61 \times 4651}$,
$E_7 = \frac{2 \times 3 \times 7 \times 88964520035219}{11 \times 13 \times 29 \times 31 \times 61 \times 4651 \times 6959}$, $E_8 = \frac{2^7 \times 3 \times 41 \times 3911 \times 121999962547}{11 \times 13 \times 17 \times 29 \times 31 \times 61 \times 113 \times 4651 \times 6959}$.

And finally, we calculate our total expected number of rolls as follows.

$$E = 1 + \sum_{i=1}^{8} P_i E_i = \frac{98081 \times 336640049 \times 20818956233}{2 \times 3^8 \times 7 \times 11 \times 13 \times 17 \times 29 \times 31 \times 61 \times 113 \times 4651 \times 6959} \approx 15.3485.$$

Solution by the Proposer

We show the expected number of rolls is about 15.3485.

This can be considered an occupancy problem with 10 indistinguishable balls placed into 6 cells. On the first roll any value can be chosen, so the cells themselves can be considered indistinguishable in that case. Later rolls must match the initial value chosen, so the cells *are* distinguishable.

To minimize the number of rolls, it is optimal to remove as many dice as possible on the first roll, so one would select any of the most frequent values rolled. The overall expected number of rolls is then the sum of the probability of rolling a maximum of n dice of some same value on the first roll, times the expected number of rolls of $10 - n$ dice to match the value removed on the first roll, as specified by the rules of the game.

Consider the second and subsequent rolls first. Let E_n be the expected number of rolls of n dice to match a given value (say 1) under the rules of the game. For one die the expected number of rolls is $E_1=1/(1/6)=6$. For two dice, the possibilities are both match a given number on the first roll (probability 1/36), one of a pair matches (probability $2 \times \frac{1}{6} \times \frac{5}{6}$) or neither matches (probability (25/36)) and we have made no progress.

Therefore $E_2 = 1/36 + 10/36(1+E_1) + 25/36(1+E_2)$ and one solves for $E_2=96/11$. Continuing in this fashion one derives the general case:

$$E_n = \frac{1}{6^n} + \frac{5^n}{6^n}(1+E_n) + \sum_{k=1}^{n-1}(1+E_n)\frac{n!}{k!(n-k)!}\frac{5^k}{6^n}.$$

This produces the table of expected number of rolls of k dice to match the value rolled on the first roll, plus the first roll itself (i.e. $(1+E_k)$ for $k = 1,\ldots,8$). We end with $k = 8$ by the pigeon hole principle (at least two dice must match on the first roll).

The next step is to calculate the probability that the first roll produces a maximum of k dice of the same value, for $k = 10, 9, \ldots 2$. This probability can be derived by examining the occupancy problem for 10 indistinguishable balls in 6 indistinguishable cells (which is related to the partition of the number 10 into 6 or less parts) and then adding up the number of rolls that contain a cell with a maximum of k items. For example, the first roll patterns that result in six dice with the same value are: {6,4,0,0,0,0}, {6,3,1,0,0,0}, {6,2,2,0,0,0}, {6,2,1,1,0,0} and {6,1,1,1,1,0}. Since the balls and cells are both indistinguishable, the number of ways these patterns can be achieved are Multinomial (6,4)×Multinomial (1,1,4) + Multinomial (6,3,1)×Multinomial (1,1,1,3) + Multinomial (6,2,2)×Multinomial (1,2,3) + Multinomial (6,2,1,1) × Multinomial (1,1,2,2)+ Multinomial (6,1,1,1,1)×Multinomial (1,4,1) = 787,500. Dividing by 6^{10} gives the probability of achieving a first roll with 6 (and no more) dice the same. Doing this calculation for all the various values and combining with the expected rolls in the first step results in the following table.

Max # The Same*	Occurrence	Probability	Expected Rolls	Expectation
10	6	0.00000010	1.0000	0.00000010
9	300	0.00000496	7.0000	0.00003473
8	6,750	0.00011163	9.7273	0.00108588
7	90,000	0.00148844	11.5554	0.01719953
6	787,500	0.01302381	12.9267	0.16835484
5	4,721,220	0.07808035	14.0237	1.09497236
4	18,774,000	0.31048764	14.9378	4.63800117
3	32,004,000	0.52928765	15.7213	8.32111190
2	4,082,400	0.06751543	16.4069	1.10772188
Total	60,466,176	1.00000000	N/A	**15.34848239**

* On First Throw

2883. Clock Progression by Marcus Upton, Harrisonburg, VA (*JRM* 38:2, p. 135)

A circular clock face has radius 12. The hour hand and minute hand have zero thickness and move continuously at a constant speed, not in discrete jumps. At 12:00 the hands coincide, of course. At how many minutes after 12:00 will they be positioned so that a disk of radius 2 can be tangent to each of them and to the circumference of the clock?

Solution by Kathleen Lewis

In the following diagram, the line segments OD and OE represent the hour and minute hands, respectively. Since the circle centered at C is tangent to both of these lines, CA is perpendicular to OD and CB is perpendicular to OE. So the triangles OAC and OBC are both right triangles having a hypotenuse of length 10 and one leg of length 2. Thus, $<AOB = 2\sin^{-1} 0.2 \approx 23.07^0$. Since the minute hand ravels 360^0 in an hour, it advances 6^0 per minute. The hour hand advances only 30^0 an hour or 0.5^0 per minute, so the two hands separate at a rate of 5.5^0 per minute. Therefore, it take about 4.2 minutes before they are in the required position.

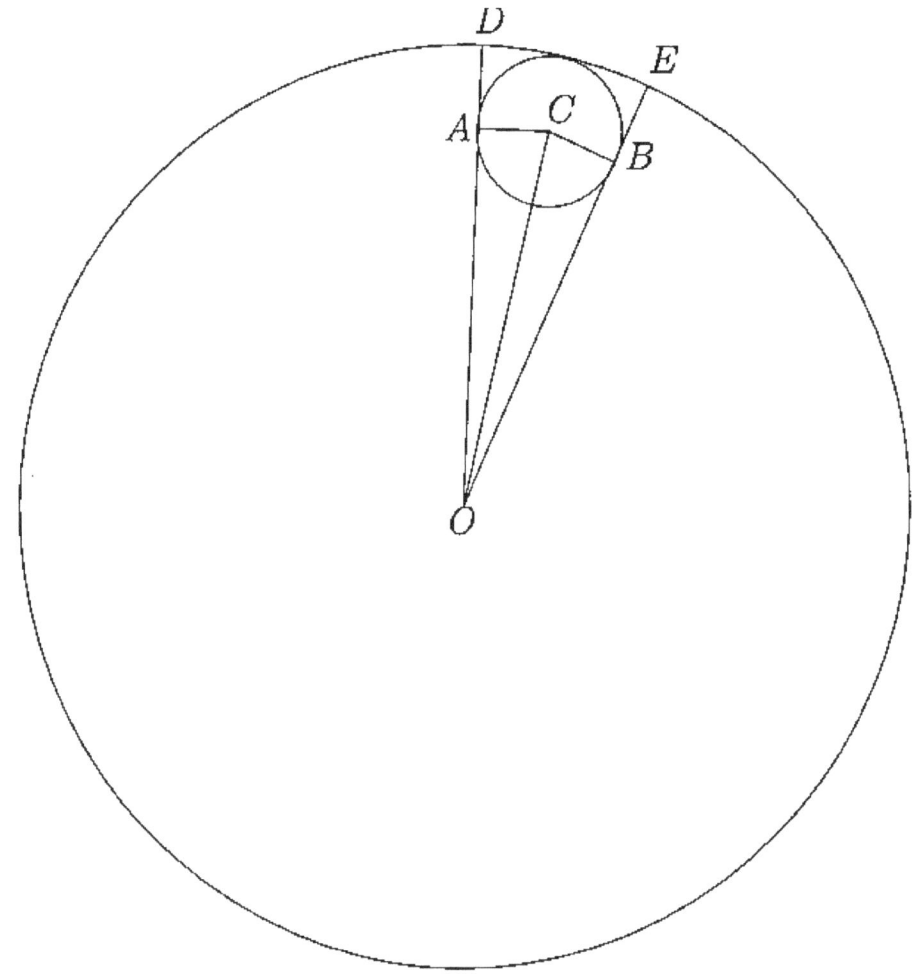

2884. Close Encounters of the *n*-th Kind by Hubert Hagadorn, Menlo Park, CA (*JRM 38*:2, p. 135)

a. Find a solution for the inequality $|x^n + y^n - z^n| < |1 + 2^n - 3^n|$, where x, y, z are distinct positive integers and n is an integer greater than three.

*b. Find another solution subject to the same conditions specified in part a.

Solution by Richard Hess

a. $(x, y, z, n) = (13, 16, 17, 5)$

b. A computer search found no other solution.

Editor's Commentary

Henry Ibstedt and **Richard Hess** both mentioned that $|7^4 + 8^4 - 9^4| = 64 = |1^4 + 2^4 - 9^4|$.

2885. Math Dice by Richard I. Hess, Rancho Palos Verdes, California (*JRM 38*:2, p. 135)

For each of the following, you may use addition, subtraction, multiplication, division, powers, decimal points, concatenation and parentheses, but no roots, factorials or other notations. Find as many solutions as possible for each part.

a. Use each of the digits 2, 3 and 4 exactly once to form an expression which equals 4.
b. Use each of the digits 2, 3 and 6 exactly once to form an expression which equals 8.

Solution by Andy Pepperdine

a. $\quad 4 = 4 \times (3-2) = 4/(3-2) = 4^{(3-2)} = \cdot 4/(\cdot 3 - \cdot 2) = 2^3 - 4 = 32^{\cdot 4}$.
b. $\quad\quad\quad\quad\quad\quad\quad\quad\quad 8 = 2^{6-3} = 32^{\cdot 6}$.

If recurring decimals are allowed, then we can also add the following.

a. $\quad 4 = \cdot \bar{4} \times 3^2 = \cdot \bar{4}/3^{-2} = \cdot \bar{4} \times (\cdot 3^{-2}) = \bar{4}/(\cdot 3^2) = \cdot \bar{4}/(\cdot 3 - \cdot \bar{2})$.
b. $\quad\quad\quad\quad\quad 8 = (2 + \cdot \bar{6}) \times 3 = (2 + \cdot \bar{6})/\cdot 3 = 36 \times \cdot \bar{2}$.

2886. Digital Fractions by Andy Pepperdine, Bath, UK (*JRM 38*:2, p. 136)

a. We can represent the number 2012 as 561348/279, using each of the digits 1 to 9 exactly once. What is the next higher number that can be so represented?
b. If we wish to find such representations using all ten decimal digits, then 2012 cannot be so represented. The next lower number with this representation is 2006=1293870/645 . Which is the next higher number with such a representation?

Solution by Henry Ibstedt
a. 2015 = 642785 / 319
b. 2015 = 1428635 / 709 = 1670435 / 829
A computer search revealed that 2033 has three different representation of either type.

2887. Prime Power Expansion by Henry Ibstedt, Issy les Moulineaux, France (*JRM 38*:2, p. 136)
a. Find the smallest prime which can be expressed in five different ways as a sum $2^n + p$, where n is an integer and p is prime.
b. Find the smallest prime which can be expressed in five different ways as a sum $2^n + p^m$, where n and m are integers, p is prime and $m > 1$ for at least one of these expansions.

Solution by the Proposer

a. $829 = 2+827 = 2^3+821 = 2^5+797 = 2^7+701 = 2^9+317$

b. $1217 = 2^2+1213 = 2^4+1201 = 2^6+1153 = 2^{10}+193 = 2^8+31^2$

2888. Determinants by Lamarr Widmer, Mechanicsburg, PA (*JRM 38*:2, p. 136)

We have $\begin{vmatrix} 1 & 2 \\ 3 & 4 \end{vmatrix} = -2$, while $\begin{vmatrix} 1 & 2 & 3 \\ 4 & 5 & 6 \\ 7 & 8 & 9 \end{vmatrix} = \begin{vmatrix} 1 & 2 & 3 & 4 \\ 5 & 6 & 7 & 8 \\ 9 & 10 & 11 & 12 \\ 13 & 14 & 15 & 16 \end{vmatrix} = \begin{vmatrix} 1 & 2 & 3 & 4 & 5 \\ 6 & 7 & 8 & 9 & 10 \\ 11 & 12 & 13 & 14 & 15 \\ 16 & 17 & 18 & 19 & 20 \\ 21 & 22 & 23 & 24 & 25 \end{vmatrix} = 0$.

What is the case for larger square matrices of this type?

Solution by Richard Hess

For all dimensions 3×3 and larger, the second row is half the sum of the first and third rows. So the rows are dependent and the determinant is 0.

2889. Pancake Flipping by Charles Ashbacher, Marion, IA (*JRM 38*:2, p. 136)

The pancake flipping problem was mentioned in a presentation by Ivars Peterson at the Iowa section meeting of the Mathematical Association of America. It has a rich history and has applications in computer network theory.

A cook is incapable of cooking a batch of pancakes where all of them are the same size. Once a stack is complete, they will be stacked in a random order and the waiter will reorganize the stack so that they go from the largest to the smallest from the bottom to the top. This is accomplished by performing a flip, which is placing a spatula somewhere in the pile and flipping the entire stack of pancakes above the spatula over.

Problem:

Assuming that it takes at most k flips to properly orient a stack of n pancakes, prove that it will take at most $k+2$ flips to properly orient a stack of $n+1$ pancakes.

Solution by Richard Hess

To arrange a stack of $n + 1$ pancakes, first find the largest, place the spatula under it and flip it and those above it so that the largest is now on top. Now flip the entire stack, leaving the largest on the bottom. So after two flips we have the largest on the bottom and by our assumption, k

flips are sufficient to correctly arrange those above it. So a total of $k + 2$ flips suffice to arrange the entire stack.

Proposers And Solvers List For Problems And Conjectures Journal of Recreational Mathematics 38(2)

P	S	Name	Location	2880	2881	2882	2883	2884	2885	2886	2887	2888	2889
■		Charles Ashbacher	Marion, IA										P
		Brian Barwell	Hampton, Middlesex, UK	S			S	S	S		S	S	S
■		Bill Dean	Wood-Ridge, New Jersey	P									
■	■	Hubert Hagadorn	Menlo Park, CA		P			P					
■	■	Richard I. Hess	Rancho Palos Verdes, CA	S	S	S	S	S	P	S	S	S	S
■	■	Henry Ibstedt	Broby, Sweden	S	S	S	S	S	S	S	P	S	S
■	■	Ken Klinger	Northbrook, IL			P	S					S	
	■	Kathleen Lewis	Brikama, Gambia					S				S	
■	■	Andy Pepperdine	Bath, UK						S	P	S		S
■		Marcus Upton	Harrisonburg, VA					P					
■	■	Proposer/ Solver	P = Proposer S = Solver										

Problems and Conjectures

Edited by: Lamarr Widmer

Readers are invited to send new problems, solutions and comments to me at *Messiah College, Suite 3041, One College Avenue, Mechanicsburg, PA 17055* or email to widmer@messiah.edu . Put each problem or solution, with your full name and postal address, on a separate sheet. Selection of solutions for publication will take place at least three months after problems appear in print.

1. Random Needle Drop by Daniel P. Shine, Cincinnati, OH

A long thin needle of length r is randomly placed on a circular table of radius r. What is the probability that the needle lies totally on the table?

2. Pythagorean Triples of Palindromes by Hubert Hagadorn, Menlo Park, CA

For Pythagorean triangles whose three sides are all palindromes, answer the following.

a. Assuming that the multiplier consists of only digits 0 and 1, find a formula for the number of triangles which are multiples of the 3-4-5 triangle such that all three side lengths have n digits.
b. What is the smallest triangle that is not a multiple of the 3-4-5 triangle?
c. Find a triangle where the common factor of the side lengths is not a palindrome.

3. Square Coverage of Equilateral Triangle by Hubert Hagadorn, Menlo Park, CA

A square and an equilateral triangle have the same area. The square is cut into three pieces so that two of these pieces will fit inside the triangle (non-overlapping) with maximum area coverage. How should the square be cut and what is the percentage of area coverage?

4. Equilateral Triangle Coverage of Square by Hubert Hagadorn, Menlo Park, CA

A square and an equilateral triangle have the same area. The triangle is cut into three pieces so that two of these pieces will fit inside the square (non-overlapping) with maximum area coverage. How should the triangle be cut and what is the percentage of area coverage?

5. A Geometric Congregation by Henry Ibstedt, Issy les Moulineaux, France

The sides of the arbitrary triangle *ABC* shown in Figure 1 are denoted *a*, *b* and *c*. R denotes the radius of the circumscribed circle and r the radius of the inscribed circle. r_a is the radius of the circle touching the side a and the extensions of the other two sides. r_b and r_c are similarly defined. In addition, *M* is the midpoint of segment *BC*, *O* is the center of the circumscribed circle, *Q* is the center of the inscribed circle and *AQ* is the bisector of the angle *A*. Prove the relation:

$$16R^2 = r^2 + r_a^2 + r_b^2 + r_c^2 + a^2 + b^2 + c^2$$

(Ed. The proposer reports that he was a subscriber to a journal in which this problem appeared in 1956. We are unfortunately unable to provide a complete citation.)

Figure 1

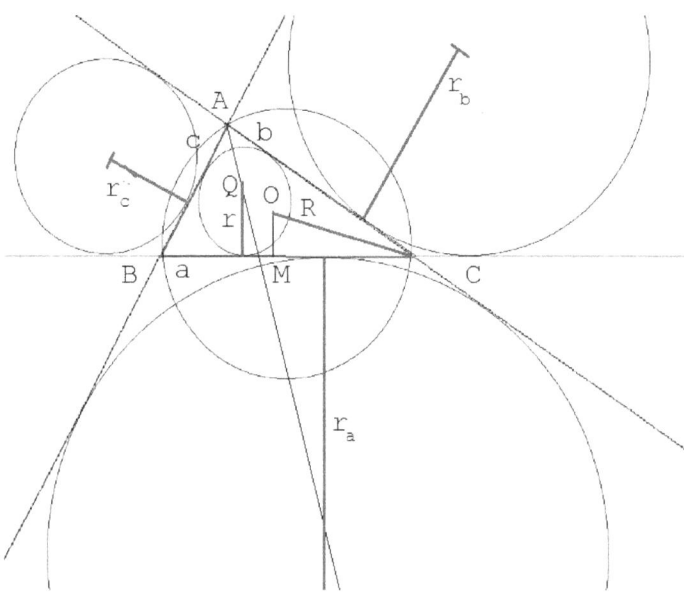

Solutions To Problems From Topics in Recreational Mathematics 3/2015

Lamarr Widmer

Messiah College
Suite 3041, One College Avenue
Mechanicsburg, PA 17055
USA

widmer@messiah.edu

1. Pythagorean Dissection by Brian Barwell, Hampton, Middlesex, UK
Find a four-piece dissection of a 7 ×7 square and a 24 × 24 square such that the pieces can be re-assembled to form a 25 × 25 square. Pieces may be turned over but the cuts may only be made along boundaries of the unit squares which make up the larger squares.

Solution by the Proposer
Figure 1 shows how to dissect the 7 × 7 and 24 × 24 squares into four pieces which can be re-assembled to make a 25 × 25 square. None of the pieces need to be turned over or rotated. This solution can be generalized to give a four-piece dissection for any Pythagorean relationship where the dimensions of the two largest squares differ by 1, i.e.
$(2n + 1)^2 + (2n^2 + 2n)^2 = (2n^2 + 2n + 1)^2$.

2. Pentomino Rectangle with Holes by Brian Barwell, Hampton, Middlesex, UK
Figure 2 shows the eighteen one-sided pentominoes and Figure 3 shows how they can be arranged to form an 8 × 15 rectangle with a central 6 × 5 hole. Use these eighteen pieces to construct a rectangle with a central rectangular hole with area greater than 30. The pentominoes may be rotated but not turned over.

Solution by the Proposer
Rectangles which might be made from the set of one-sided pentominoes and which contain holes larger than 30 squares include the 9 × 14 rectangles with a 3 × 12 hole and the 11 × 12 rectangle with a 7 × 6 hole. I was unable to find a solution in either of these cases. I was able to find a solution shown in Figure 4 for the 9 × 15 rectangle with a 5 × 9 hole.

Figure 1

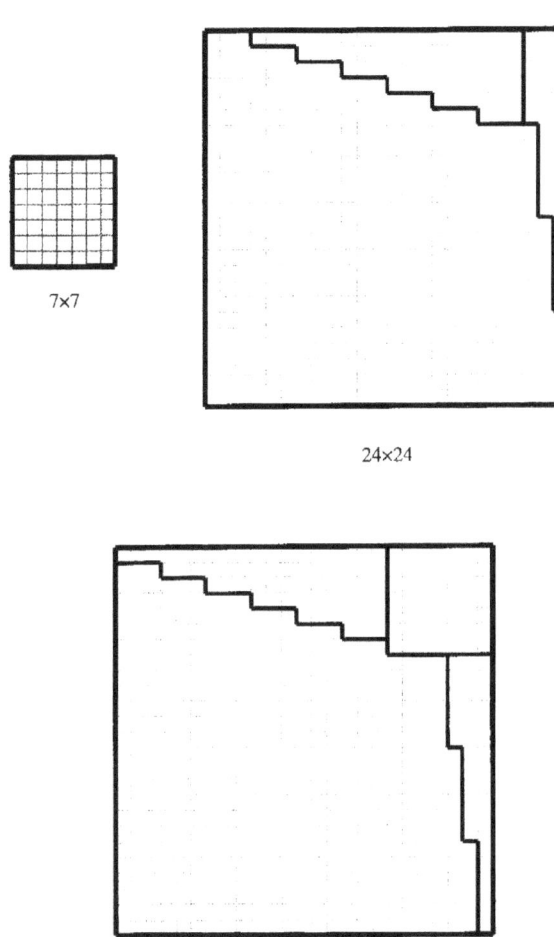

7×7

24×24

25×25

Figure 2

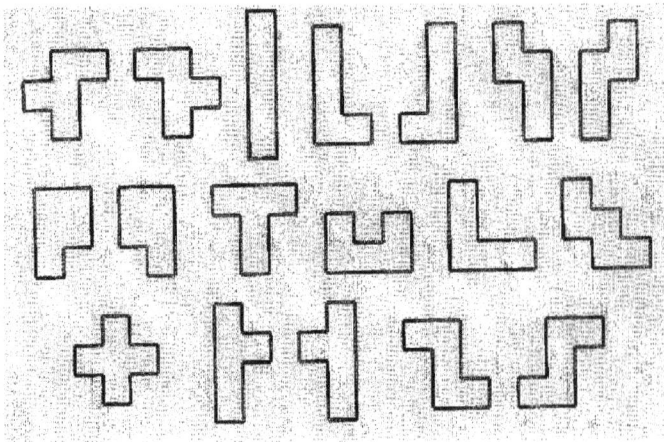

Figure 3

Figure 4

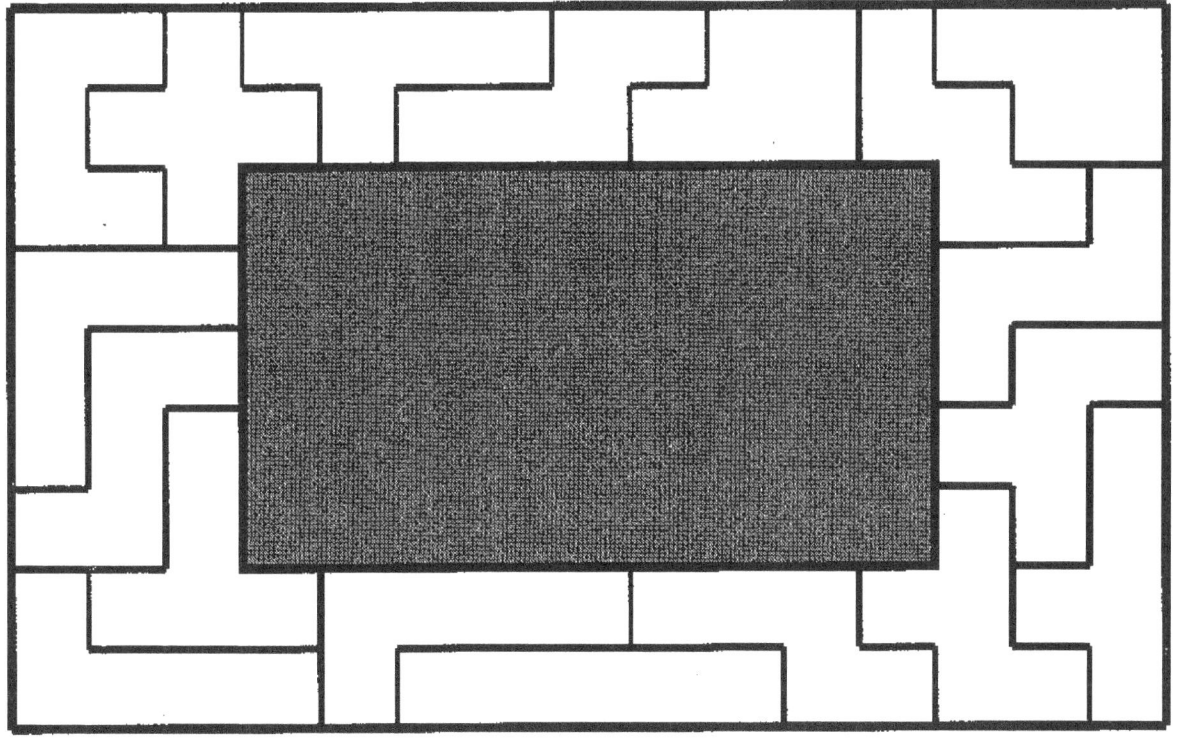

3. Wall Scraper by Hubert Hagadorn, Menlo Park, CA

A semicircle of diameter 2 is able to move along a path of unit width having a sharp right angle turn, sliding, rotating and then sliding again as shown in Figure 5. What is the shape of largest area that is able to travel along this path and negotiate the turn?

Figure 5

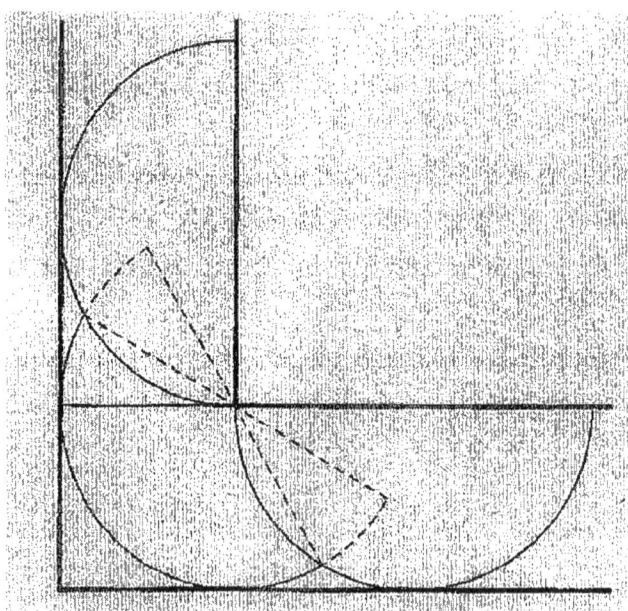

Solution by the Proposer

Not known. The area of the semicircle is about 1.571, but larger areas are also possible. Figure 6 shows the right side of a symmetrical shape for a region having an area of 2.218. Areas were in agreement to four decimal places for calculations based on ellipse formulas, and those based on calculation at about 4170 discrete points along the perimeter.

Figure 6

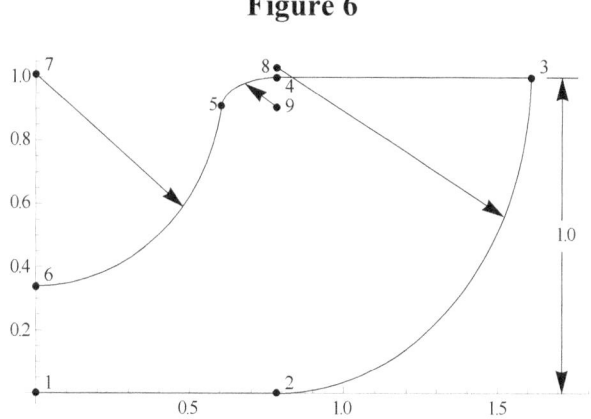

The region is defined by two lines and arcs of three ellipses. Horizontal lines are from point 1 to 2, and from point 3 to 4. The centers for ellipses 1, 2 and 3, are at points 7, 8, and 9, respectively. The slopes of ellipses 1, 2, and 3 are zero at points 6, 2, and 8, respectively. Points 2, 4, 8, and 9 have the same x-coordinate. Ellipses 1 and 3 are tangent at point 5. In Figure 7 the region is shown at four positions as it rotates from zero degrees to 45 degrees in 15 degree increments.

Figure 7

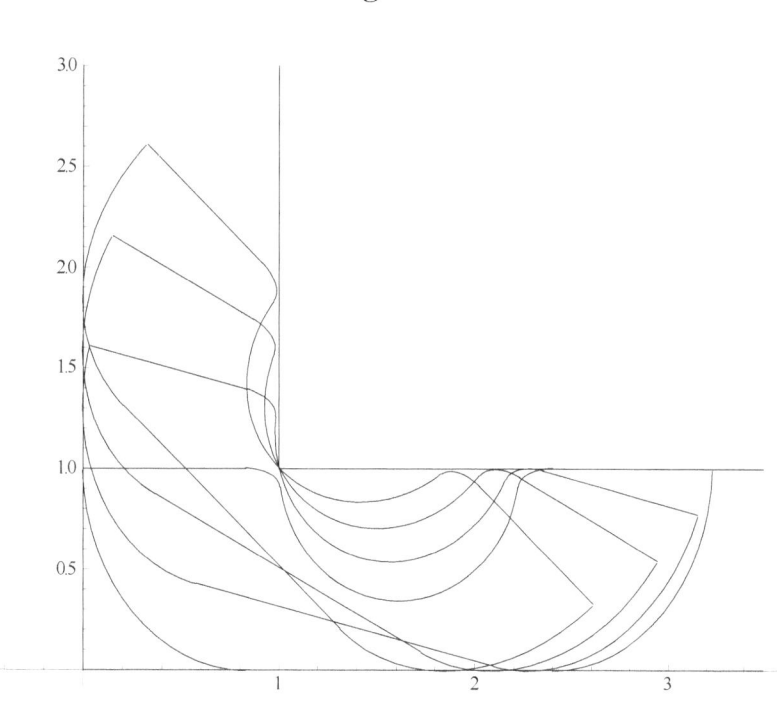

Table 1 gives the semi-axes of the three ellipses, where a and b refer to their axes aligned with the x and y axes, respectively. Table 2 gives the coordinates of the nine points. The seven values with asterisks are those parameters optimized. Other values excluding 0 and 1 are derived values.

Table 1

Ellipse	a	b
1	*0.612140	*0.670108
2	*0.830403	*1.031406
3	0.177115	*0.094859

Table 2

Point	x	y
1	0	0
2	0.78233	0
3	1.61235	1
4	0.78233	1
5	0.605669	0.911926
6	0	0.339000
7	0	*1.009108
8	0.78233	1.031406
9	0.78233	*0.905141

4. Point of Concurrency in a Square by Subramanyam Durbha, Norristown, PA

Let $ABCD$ be a square. Let E be the midpoint of BC, F be the midpoint of CD and G the midpoint of BE. Prove that the lines AE, BF and DG are concurrent.

Solution by Andy Pepperdine

In Figure 8, $ABCD$ is the square, E and F are midpoints of sides BC and CD, and G the midpoint of BE. Let X be the intersection of AE and BF, and draw the lines XG and XD. Then to show that DG passes through X is the same as to show that opposite angles $\angle GXB$ and $\angle DXF$ are equal. Let $\angle BAE$ be θ. Then since triangles BAE and CBF are congruent, $\angle CBF$ is also θ, and as triangles BAE and XBE are similar, AE and BF are perpendicular, and G is the circumcenter of triangle XBE, so that $GX = GB$. Triangle BGX is isosceles and so $\angle GXB$ is also θ. Draw the line AF. Triangles BAE and DAF are also congruent and hence $\angle DAF$ is θ. But $\angle ADF$ and $\angle AXF$ are both right angles, and so $ADFX$ is a cyclic quadrilateral, and hence $\angle DXF = \angle DAF = \theta$. So $\angle GXB = \angle DXF$ and the desired result is proved.

Figure 8

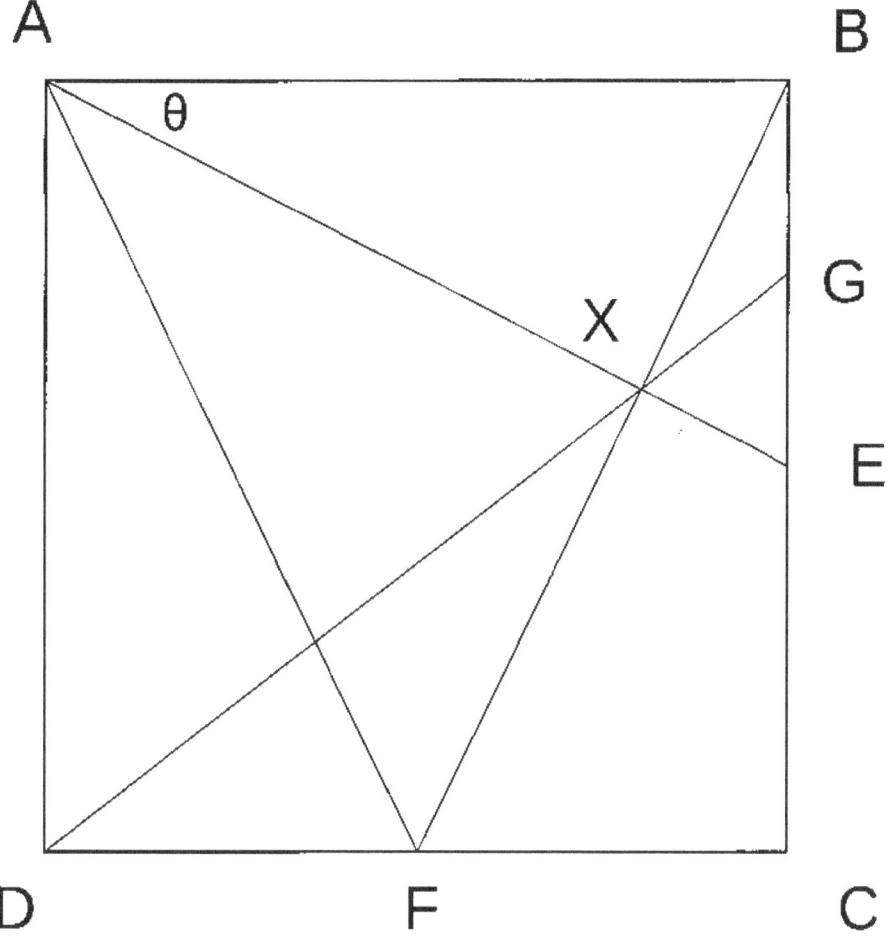

In the Loop—the Mini and Max of Paths

Kate Jones

Forming connections, paths and passages is a venerable genre in games, tracing back to ancient labyrinths. Pursuing our theme that games are microcosms of the real world, enacting on gameboards the dynamics of social order, we can see linkages as forming walls, borders, roads, bridges and enclosures, symbolic of territorial conquests, property holdings, even map making.

Loops especially suggest the recurring mathematical patterns and cycles of life, of closure, of safety as well as traps. Paths and loops do, in fact, mirror systems in nature, like tree roots and branches, vines, rivers and canyons, spiderwebs, arteries and cells, intestines, and even the convoluted structures of our brains.

Let's take a look at several original path-based sets that serve for puzzles whose tiles can form both tiny bubbles and full-length loops—the mini and max. Most were introduced by Kadon between 1984 and 2015. All are still in print.

Kaliko

In 1968, an unusual game was published by KMS Industries under the name "Psyche-Paths," consisting of 85 hexagonal tiles with three curvy paths on each, connecting pairs of sides in five different path patterns. Three colors occurred on the path sections in every possible combination. The tiles' path segments connected by similar colors and formed meandering loops and networks. The inventors were two university professors, Chuck Titus and Craige Schensted (changed later to Ea Ea) of Ann Arbor, Michigan.

From 1982-1987, the set was published by Future Classics under a new name, KALIKO, in screen-printed clear acrylic. In 1990 Kadon received an exclusive license to produce KALIKO, launched in 1991 in lasercut acrylic and changed to lasercut wood in 2001. The set lends itself to the most beautiful puzzle loops and symmetries (left) and the challenge of forming very long loops (right).

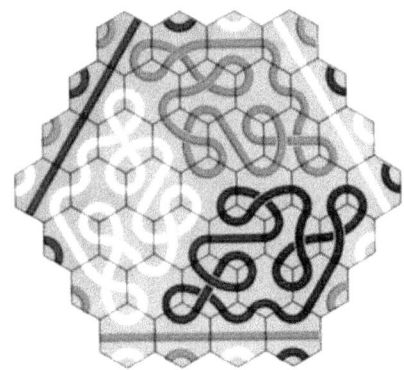 

Octiles

In 1983, *Omni* magazine published "The World's Hardest IQ Test." One of the questions on the test was so clever and beautiful that Kate Jones (of Kadon Enterprises, Inc.) decided it would have to be the basis of a game someday. The question had a diagram of 16 circles with 4 arcs each in different conn ectivities, and it asked: "The missing pattern is ___ ." Sketching out the answer, Kate filed it for future reference.

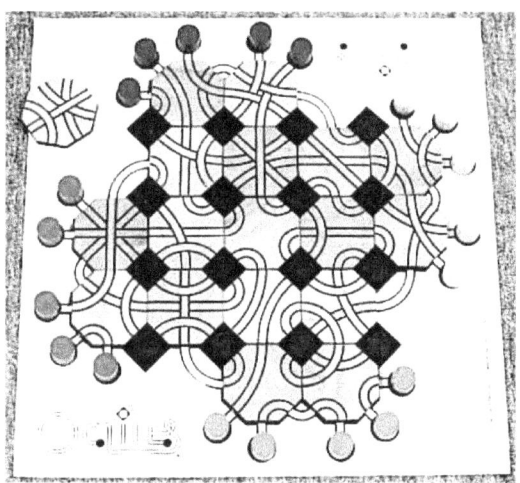

A few months later, Dale Walton, the son of a Kadon customer, visited to show Kate a new game idea. Dale's paper model of 18 cut-out octagons with arcs connecting their sides essentially had the *Omni* puzzle patterns! The extra two tiles were mirror opposites of the "missing pattern"—the asymmetrical one. Each tile was unique, and new connections would alter on every turn where paths would go, in astronomical variety.

Kadon published OCTILES in 1984 with a roll-up vinyl gameboard (above). The deluxe wood format (below) was introduced in 2009, on the occasion of the game's 25th anniversary.

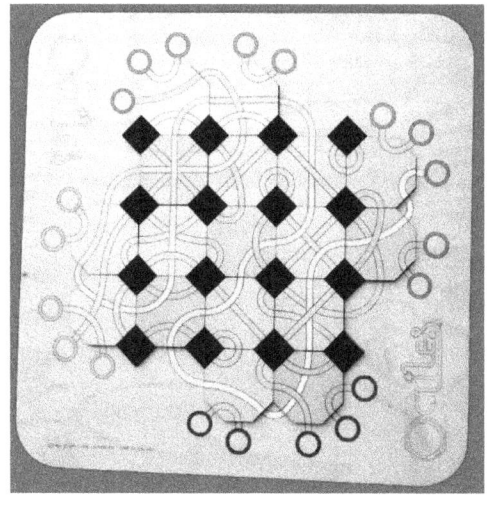

One of the many solitaire challenges for OCTILES asks for an arrangement of a single path to traverse every tile and a maximum of ramps. Can you improve on the solution shown at right? Alternately, how many separate closed loops can be formed?

Dezign-8

The forerunner of DEZIGN-8 (below) was a small wood set of only 30 tiles created by Bill Briggs, an architect, in 1959. It languished until 1999, when his widow, Eloise Briggs, contacted Kate Jones about marketing the set, with rules for a smart game of strategy for up to 4 players.

Kate licensed the game and developed it further, adding 4 missing tile patterns to give the set all the possible combinations of 1 to 4 paths transiting and exiting the tiles. Here are the 8 topologically distinct tile patterns of DEZIGN-8:

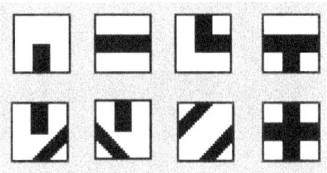

Research into how many subgroups and inner loops could be formed brought a surprise: the number of groups always equalled the number of loops, from 1 to 19. The figure below shows the one enclosed inner space, the loop within the one continuously connected circuit.

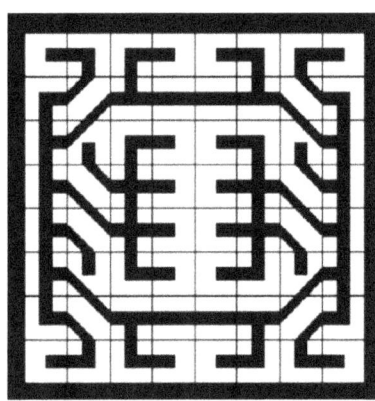

The total number of solutions for achieving one loop is not known. At the other extreme is the 19-group solution and its 19 loops (above).

Kadon released DEZIGN-8 in 2000. In 2001 it was included on *Games Magazine*'s list of the 100 best games of the year. If only Bill Briggs could have seen the success of his brainchild from four decades before.

QUINTAPATHS

Another loop-forming set of early vintage is QUINTAPATHS, invented by Scott Kim in 1969, when he was still in high school. It was published by Kadon in 1999. The set has 20 uniform 1x5 sticks, each with a different number or arrangement of black squares on top, from 0 to 5:

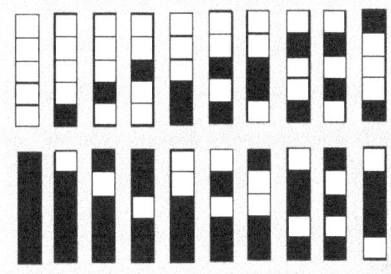

These can make a single-track black loop in their 10x10 frame (left). And the same 20 tiles can maximize the number of single enclosed spaces ("islands") as in this *yin-yang* array (right).

 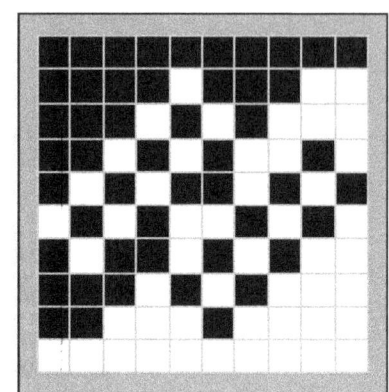

Arc Angles

For Kadon's 25th Anniversary in 2004, Kate designed a special puzzle with a theme of 25. Five rounded-off "kites" form a ring, and the five rings total 25 pieces. All-different paths on five levels step up or down to connect the five marks on their edges—arcs with angles.

All five rings match in symmetrical loops. One complete 25-piece loop can take a surprising variety of shapes, from largest (far right) to tightest enclosed area. As the tiles join in waves, not just rings, getting the loop to close is no mean feat. A prize awaits the smallest enclosure.

 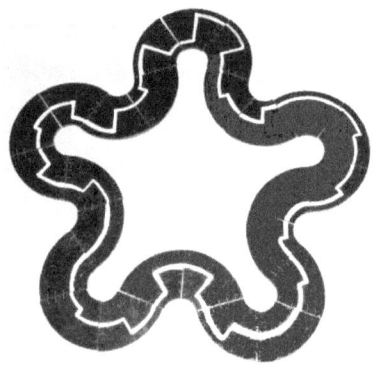

Escape The Plague

Another special occasion that required a custom design was the 34th International Puzzle Party held in London in August 2014. For this event, Kate designed a humorously macabre maze puzzle, "London 1665: ESCAPE THE PLAGUE", with 16 square wood tiles and 4 movable outer wall strips with connecting paths, presented in a small coffin box (below).

The challenge was to select tile patterns, as many as possible symmetrical, that would allow assembly with one single path to go from a start point on the interior to the escape point through the outer wall and cover exactly once every line segment on the tiles and on the wall strips. Here is one such solution:

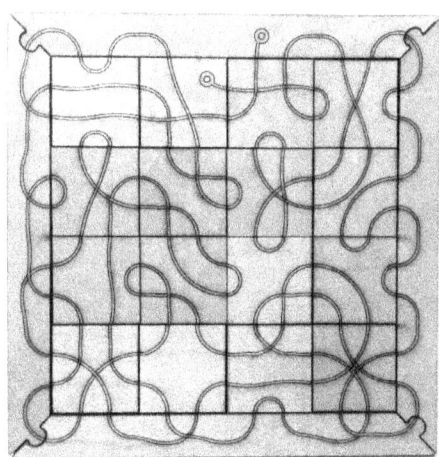

A preliminary search by computer has established that ESCAPE THE PLAGUE has literally millions of solutions to assemble a single path, including the 6 distinct ways of hooking the outer wall strips together.

The logical next question is: What is the maximum number of *separate* loops that can be formed? The instruction booklet showed a 16 (below left). Several puzzle solvers have topped that, and the best result at the time of this writing is the 20 (below right) found by George Sicherman. The reader is invited to improve on that or prove it maximum.

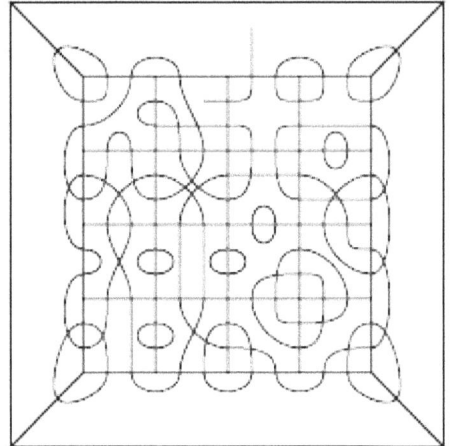

©2016 Kadon Enterprises, Inc. *All game names are proprietary trademarks of Kadon Enterprises, Inc. You can see these puzzles and many more, most of them suitable for ages 8 to adult, on Kadon's website,* **www.gamepuzzles.com**

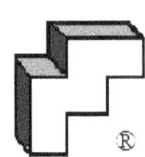

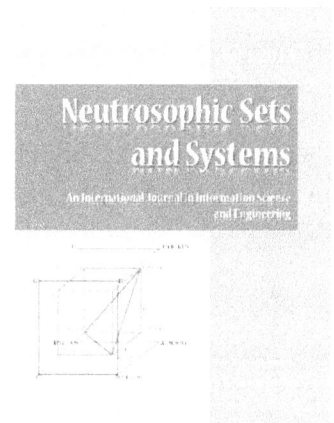

Editor-in-Chief:

Prof. Florentin Smarandache

Department of Mathematics and Science

University of New Mexico

705 Gurley Avenue

Gallup, NM 87301, USA

E-mail: smarand@unm.edu

Home page:
http://fs.gallup.unm.edu/NSS

Associate Editors:

Dmitri Rabounski and Larissa Borissova, independent researchers.

Said Broumi, Univ. of Hassan II Mohammedia, Casablanca, Morocco.

A. A. Salama, Faculty of Science, Port Said University, Egypt.

Yanhui Guo, School of Science, St. Thomas University, Miami, USA.

Francisco Gallego Lupiañez, Universidad Complutense, Madrid, Spain.

Peide Liu, Shandong Universituy of Finance and Economics, China.

Pabitra Kumar Maji, Math Department, K. N. University, WB, India.

S. A. Albolwi, King Abdulaziz Univ., Jeddah, Saudi Arabia.

Mohamed Eisa, Dept. of Computer Science, Port Said Univ., Egypt.

Neutrosophic Sets and Systems has been created for publications on advanced studies in neutrosophy, neutrosophic set, neutrosophic logic, neutrosophic probability, neutrosophic statistics that started in 1995 and their applications in any field, such as the neutrosophic structures developed in algebra, geometry, topology, etc.

The submitted papers should be professional, in good English, containing a brief review of a problem and obtained results. Neutrosophy is a new branch of philosophy that studies the origin, nature, and scope of neutralities, as well as their interactions with different ideational spectra.

This theory considers every notion or idea <A> together with its opposite or negation <antiA> and with their spectrum of neutralities <neutA> in between them (i.e. notions or ideas supporting neither <A> nor <antiA>). The <neutA> and <antiA> ideas together are referred to as <nonA>.

Neutrosophic Set and Logic are generalizations of the fuzzy set and respectively fuzzy logic (especially of intuitionistic fuzzy set and respectively intuitionistic fuzzy logic). In neutrosophic logic a proposition has a degree of truth (*T*), a degree of indeterminacy (*I*), and a degree of falsity (*F*), where *T, I, F* are standard or non-standard subsets of $]^{-}0, 1^{+}[$.

Neutrosophic Probability is a generalization of the classical probability and imprecise probability. Neutrosophic Statistics is a generalization of the classical statistics.

What distinguishes the neutrosophics from other fields is the <neutA>, which means neither <A> nor <antiA>. <neutA>, which of course depends on <A>, can be indeterminacy, neutrality, tie game, unknown, contradiction, ignorance, imprecision, etc.

All submissions should be designed in MS Word format using our template file:

http://fs.gallup.unm.edu/NSS/NSS-paper-template.doc

A variety of scientific books in many languages can be downloaded freely from the Digital Library of Science:

http://fs.gallup.unm.edu/eBooks-otherformats.htm

To submit a paper, mail the file to the Editor-in-Chief. To order printed issues, contact the Editor-in-Chief. This journal is non-commercial, academic edition. It is printed from private donations.

Information about the neutrosophics you get from the UNM website:

http://fs.gallup.unm.edu/neutrosophy.htm

The home page of the journal is accessed on

http://fs.gallup.unm.edu/NSS

BOOKS IN RECREATIONAL MATHEMATICS BY CHARLES ASHBACHER AND ASSOCIATES

Topics in Recreational Mathematics 1/2015 ISBN 978-1507603215

Topics in Recreational Mathematics 2/2015 ISBN 978-1508617099

Topics in Recreational Mathematics 3/2015 ISBN 978-1511641005

Topics in Recreational Mathematics 4/2015 ISBN 978-1514317518

Topics in Recreational Mathastics 5/2015 ISBN 978-1519115676

Topics in Recreational Mathematics 1/2016 ISBN 978-1530003655

Topics in Recreational Mathematics 2/2016 ISBN 978-1534964846

Alphametics as Expressed in Recreational Mathematics Magazine ISBN 978-1508538134

Ten Year Cumulative Index to the Journal of Recreational Mathematics, edited by Joseph S. Madachy and Charles Ashbacher ISBN 978-1508936800

Alphametics Expressing Thoughts From the Star Trek Original Series ISBN 978-1512152784

Mathematical Cartoons ISBN 978-1514207130

Solved Problems in Statistical Inference ISBN 978-1515215622

Associates

Artist Caytie Ribble

Editor Rachel Pollari

Editor Jennifer Corrigan

Artist Jenna Richardson

www.ingramcontent.com/pod-product-compliance
Lightning Source LLC
Chambersburg PA
CBHW080705190526
45169CB00006B/2246